Renewable Energy Integrated Seawater Desalination Technologies

This book bridges the gap between conventional processes and renewable energy procedures by offering a comprehensive introduction to seawater desalination technologies.

With a special focus on water quality, plant design, and the health and environmental impacts of desalination, it provides innovative solutions to optimize facilities for better accessibility to desalinated water. The first part provides a technical overview of seawater quality and desalination procedures, including thermal desalination, membrane desalination, and alternative processes like desalination batteries and capacitive deionization. The second part introduces novel freshwater production technologies using renewable energy, such as solar, wind, and geothermal desalination. With fresh insights from experts in the field, the book's value lies in providing valuable, accessible, and up-to-date knowledge about desalination to readers.

This text will be of significant interest to professionals in water resource management, and academic researchers in desalination technology and environmental engineering.

Renewable Energy Integrated Seawater Desalination Technologies

Manan Shah
and
Mitul H. Prajapati

CRC Press
Taylor & Francis Group
Boca Raton London New York

CRC Press is an imprint of the
Taylor & Francis Group, an **informa** business

Designed Cover Images: Shutterstock

First published 2025
by CRC Press/Balkema
4 Park Square, Milton Park, Abingdon, Oxon, OX14 4RN

and by CRC Press/Balkema
2385 NW Executive Center Drive, Suite 320, Boca Raton FL 33431

CRC Press/Balkema is an imprint of the Taylor & Francis Group, an informa business

British Library Cataloguing-in-Publication Data
A catalog record for this book is available from the British Library

ISBN: 978-1-032-87089-2 (hbk)
ISBN: 978-1-032-87090-8 (pbk)
ISBN: 978-1-003-53089-3 (ebk)

DOI: 10.1201/9781003530893

Typeset in Times
by SPi Technologies India Pvt Ltd (Straive)

Contents

Preface

The concept of a novel book on *Renewable Energy Integrated Seawater Desalination Technologies* has been in the minds of the authors for a long time. Their personal experiences living in Gujarat, India, a region rich in oceanic resources and renewable energy, have deeply influenced their understanding and passion for this field.

This book is a pioneering effort to bridge the gap between conventional processes and sustainable renewable energy procedures in the field of seawater desalination. It aims to offer a comprehensive, concise introduction to these technologies, paving the way for a more sustainable future.

The first part of this book provides a technical but reader-friendly summary of seawater quality and an overview of desalination procedures, ranging from thermal to membrane processes. It emphasizes key technical aspects and future development prospects. It also provides a quick overview of the economics of producing freshwater from seawater.

The book's second part presents innovative freshwater production systems that utilize renewable energy sources. Initially, solar energy-integrated desalination technologies replicated the natural water cycle; however, advancements have led to the development of complex systems that use geothermal, wind, photovoltaic, and other sources to power traditional and innovative desalination methods. Desalination batteries and capacitive deionization represent emerging technologies that promise to revolutionize freshwater production. These systems not only offer improved efficiency but also present opportunities for energy recovery and storage, thus aligning with the principles of sustainable development. While most of these solutions remain in preliminary development phases, only some are recognized as successful examples in remote regions that effectively address water scarcity challenges.

The book's final section describes drinking water quality, technology, and environmental issues to help both suggested and existing desalination facilities be optimized so that countries and users can benefit from better accessibility to desalinated water while maintaining quality safety and environmental protection. Apart from the quality and safety of the produced drinking water, several other health and environmental issues, such as energy conservation and sustainability, become apparent when evaluating the effects of desalination operations. They could also apply to any major development project in a coastal area.

Collectively, these chapters offer a thorough and interdisciplinary perspective on desalination methods. This book is a collection of facts and a platform for engaging in the ongoing discussion on sustainable water management. Incorporating sustainable energy sources and addressing environmental and health considerations invites readers to be part of the solution for future water challenges.

The book offers a comprehensive and systematic description of various seawater desalination and plant specifications. It provides practical insights into the technological benefits and drawbacks, making it a valuable resource for engineers, academics, researchers, technicians, managers, private and public institutions, students, and citizens interested in desalination and sustainability.

About the Authors

Manan Shah holds a B.E. in Chemical Engineering from L.D. College of Engineering, Gujarat, India, and an M. Tech. in Petroleum Engineering from the Department of Petroleum Engineering at Pandit Deendayal Energy University (PDEU), Gujarat, India. He completed his Ph.D. in Geothermal Energy at PDEU. He is currently an assistant professor in the Department of Chemical Engineering at the School of Energy Technology (SOET), PDEU. Dr. Shah has research expertise in the field of Machine Learning (ML) and Deep Learning (DL) applications in the domains of energy, medicine, chemical, and environment. Dr. Shah possesses both experimental and computation expertise and has carried out multiple projects in the mentioned fields. In the ML domain, he has worked with various time series and non-time series datasets to predict air quality, water quality, diabetes detection, and cancer detection projects. Dr. Shah has developed various ML and DL algorithms which can be applicable to various applications. He has published extensively in prestigious international journals, contributing valuable insights to multiple sectors.

Mitul H. Prajapati, an assistant professor in the Department of Chemical Engineering at S.S. Agrawal College, Gujarat, India, and is also a gold medalist in M.E. Chemical Engineering with a specialization in Computer-Aided Process Design from L.D. College of Engineering, Ahmedabad under Gujarat Technological University (GTU). With over five years of academic and research experience, his expertise spans renewable energy, environmental sustainability, and process equipment design. His research focuses on groundwater quality, desalination systems, geothermal energy, sustainable agriculture, and soil health monitoring. He has published numerous high-impact papers in renowned journals. He holds three patents on geothermal and solar integrated desalination and hydrogen systems, showcasing practical solutions for India's energy and water challenges.

Water Quality 1

1.1 INTRODUCTION

Life depends on water; hence, everyone needs access to a sufficient (safe, accessible) supply (Yvana, 2021). Increasing the availability of clean drinking water can have a real positive impact on health. The goal should be to ensure safe drinking water quality. Whether used for irrigation or drinking, water quality significantly impacts health in developing and developed countries (Kumar et al., 2018). To address the problem of raising water quality, nations create norms to ensure public health. Understanding this, the WHO has made several standard "guidelines" that offer a reliable evaluation of the health risks connected to exposure to health hazards by water and various methods for their management (WHO, 2011).

Besides the need for water to sustain physical activities, people also utilize it for household activities, including meal preparation, clothing washing, and hygiene (Fewtrell, 2013). Humans have been consuming various water creatures and plants for a long time. Early human communities grew up where lakes or streams provided a steady water supply. Humans eventually discovered how to transport, store, and irrigate crops and access subsurface water supplies (Graham, 1999). Humans were then able to establish formerly unsuitable and arid regions. The availability of water determines how a region's population grows even now. Water is an essential industry component for power generation, which is directly using the energy of flowing water to move water wheels or turbines, steam generating, cooling, and processing (Vigil, 2023). Water may be a reagent, solvent, or component in processing. It can also transport or wash materials, and processing wastes are frequently dumped in water (Singh et al., 2019).

The spatial distribution of vegetation across the Earth's surface is primarily determined by the presence and accessibility of water, surpassing the influence of any other factor (Ghenai et al., 2021). Lush flora thrives in well-hydrated places, but arid regions lack plentiful plant life (Boyd, 2000). Water significantly influences the formation of the Earth's surface utilizing dissolution, erosion, and deposition. Significant water bodies substantially influence the air temperature of the land masses surrounding them (Prajapati et al., 2021). This phenomenon is particularly evident in coastal regions where

DOI: 10.1201/9781003530893-1

the climate may be cooler than anticipated due to a cold ocean current nearby or, conversely, where the climate may be warmer than expected due to the influence of a warm ocean current. Therefore, ecosystems rely heavily on water (Hu et al., 2013).

At the same time, as people became aware of how to exercise some control over the amount of water available to them, they discovered that different types of water had varying characteristics, such as temperature, color, taste, and odor (Sarda & Sadgir, 2015). They observed that these characteristics affected the appropriateness of water for particular applications. Consumption of salty water by humans and livestock and irrigation was neither possible nor desirable. It was determined that clear water was superior to turbid water for usage in the home. It is possible that humans could become unwell or perhaps die if they consume certain types of water (Panagopoulos, 2021). The notions of water quantity and water quality were likely formed at the same time. However, throughout the majority of human history, there were very few methods for measuring water quality that went beyond sensory experience and observations of the impacts that particular waters had on living creatures or other water applications (Holland, 2005).

The term "water quality" refers to the degree to which water is suitable for a specific application, and a water quality variable is any physical, chemical, or biological characteristic that impacts whether or not water is suitable for use by humans or by natural ecological systems (Yadav et al., 2022). In order to serve as guides for choosing water supplies for a variety of activities or for preserving water bodies from pollution, standards for water quality have been implemented. When it comes to health, the quality of the water that is consumed is vital (Darre & Toor, 2018). The water consumed must not have any disease-causing organisms, high amounts of minerals, or poisons; it must not contain any of these characteristics. When it comes to drinking water, most people prefer it to be clear and free of unpleasant odors or tastes (Fewtrell, 2013). In addition, standards for water quality are developed for waters used for bathing and recreational purposes, as well as waters used for cultivating or catching shellfish. As a result of coming into contact with contaminated water, diseases might be transmitted. Several shellfish, including oysters, can potentially acquire infections or harmful substances from the sea, making them unsafe for human eating (Omer, 1975).

Water quality for livestock does not need to be as high as water for domestic use, but it must not be harmful or toxic to animals (Boyd, 2000). High levels of minerals in irrigation water can harm plants through osmotic stress, and the water used for irrigation should not contain any hazardous compounds to plants. Water used in industrial processes must also meet the necessary quality standards (Koroneos & Roumbas, 2012). For specific procedures, it is necessary to have exceptionally high-quality water. Even the water used for boiler feed must not have an excessive amount of suspended particulates or a high level of carbonate hardness. Particles can accumulate in plumbing systems, and calcium carbonate can solidify and create deposits known as scale. Acidic waters can potentially induce significant corrosion on metal things they come into contact with (Graham, 1999).

Water is classified into different categories based on its intended usage (Darre & Toor, 2018). The initial water quality standard is established to ensure safe drinking, domestic usage, and other industrial applications. The salinity of this water group ranges from 5000 to 1000 ppm (Vigil, 2023). This variety of water is commonly located in natural bodies such as lakes and rivers and can also be produced using industrial desalination methods (Kucera, 2014). Urban areas often use water with a salinity level below 150 ppm for human consumption, whereas water with a higher salinity level, up to 1000 ppm, is used for various household purposes (Ghenai et al., 2021). This method becomes more effective when people limit themselves to no more than 2 L of water each day, which has a low salinity of 150 ppm. Conversely, the average individual consumption rate for various household activities, such as washing, cooking, gardening, cleaning, and other uses, ranges from 200 to 400 l/day (Vigil, 2023).

Boiler makeup water and water used in electronic and pharmaceutical applications set the bar for industrial water quality (Chhaya & Prajapati, 2020). A maximum salinity of 5 ppm is allowed for this application's water quality. Ion exchangers, which can process water from rivers or desalinization plants with low salinity, are essential for this exceptionally high purity level. Water quality standards for boilers are stricter than those used for other industrial uses (Dhok et al., 2011). Chemical processes, food and dairy, cleaning, and refrigeration are among its many uses. A salinity range of 1000–3000 ppm is used to classify the second type of water. Irrigation and industrial cooling are two good uses for this water type (Amaaliya & Kumar, 2013). Waters with a higher salinity, such as brackish and seawater, fall under this category. Bacterial saltwater typically has a salinity level between 3000 and 10,000 parts per million (Graham, 1999). The average salinity of seawater is 34,000 parts per million. Water is considered high salinity if it is more than 10,000 parts per million (Koroneos & Roumbas, 2012). Ambient and topographical factors influence the salinity of saltwater, which varies depending on local conditions. For instance, the salinity in enclosed seas is higher than in open oceans and seas. In addition, the degree of salinity is likely to be higher in oceans located in hotter regions or those that drain saltwater rapidly (Jarin et al., 2023).

The chapter's comprehensive exploration of water quality underscores the fundamental significance of ensuring access to clean and safe drinking water for human health and well-being. From historical perspectives to modern-day challenges, the study emphasizes the critical role of water quality in various aspects of life, including domestic use, agriculture, and recreational activities. Establishing standards for water quality serves as a guiding framework to safeguard human health and ecological systems from potential pollutants and contaminants. The study also sheds light on the intricate dynamics of water quality management in coastal zones, where the delicate balance between natural processes and human activities necessitates careful monitoring and regulation. The chapter includes the following sections: Section 1.2 delves into the vast expanse of Earth's waters, where 97.5% is saline. This section explores the intricate balance of sodium, calcium, and chloride ions that constitute the ocean's salinity. Section 1.3 discusses water quality parameters affecting desalination. It highlights the importance of understanding the various inorganic ions and other compounds in the source water for desalination, which can be marine, brackish, or highly mineralized groundwater. The section emphasizes the need to reduce or eliminate these compounds to ensure the treated water is safe for consumption and other uses. Section 1.4 delves into the characteristics of brackish groundwater, which is defined by its total dissolved solids (TDS) content. This section explores the classification of water based on TDS levels, distinguishing brackish water from fresh and seawater. It highlights the variability of brackish water's TDS and major ion constituents, which are influenced by geographic location, local geology, and hydrologic conditions. Coastal water quality management is covered in Section 1.5. Highlighting its usage in recreational activities, it highlights the importance of seawater as a crucial natural resource in coastal tourist zones. Because of the interplay between economic, environmental, and anthropogenic processes, managing resources in coastal zones is complicated and frequently results in disputes over resource use.

1.2 SEAWATER COMPOSITION

The amount of water that makes up the Earth is around 1.4×10^9 km^3, which accounts for approximately 70% of the planet's surface area. Approximately 97.5% of this vast quantity consists of saltwater (Holland, 2005). In contrast, only 2.5% of the overall water supply is freshwater, and the majority of this amount, 80%, is either frozen in ice caps or mixed with soil moisture. For human consumption, neither of these forms is readily available (Gulamhussein et al., 2023). It is believed that the remaining quantity, approximately 0.5%, is sufficient to sustain all forms of life on Earth (Ludwig, 2022). This water is not distributed uniformly throughout the plant and is not there when or where it is needed. This is a very unfortunate situation (Shaffer et al., 2012). Bays, estuaries, and seas are partially isolated bodies of saltwater subject to significant variability due to the effects of freshwater flows and evaporation. In contrast, the open ocean experiences seawater quality variations of no more than 10% over time and place (Khawaji et al., 2008) (Table 1.1).

The primary ions present in seawater are magnesium (Mg^{2+}), sodium (Na^+), sulfate (SO_4^{2-}), potassium (K^+), calcium (Ca^{2+}), and chloride (Cl^-) (Holland, 2005). However, seawater contains all other ions that exist in nature, usually at somewhat lower concentrations (Miller et al., 2010; Rajkamal et al., 2016). There is no change in the

TABLE 1.1 Seawater composition (El-Dessouky & Ettouney, 2002)

Compound	*COMPOSITION*	*MASS PERCENT*	*PPM*
Chloride	Cl^-	55.03	19810.8
Magnesium	Mg^{+2}	3.69	1328.4
Sulfate	$(SO_4)^{-2}$	7.68	2764.8
Sodium	Na^+	30.61	11019.6
Potassium	K^+	1.16	417.6
Calcium	Ca^{+2}	1.16	417.6
Bromine	Br^-	0.19	68.4
Carbonic acid	$(CO_3)^-$	0.41	147.6
Boric acid	$H_3BO_3^-$	0.07	25.2
Total		**100**	**36,000**

TABLE 1.2 Fresh and seawater thermodynamic characteristics at 25°C

Thermodynamic property	*SEAWATER*	*FRESHWATER*
Density [kg/m^3]	1023.8	997
Viscosity [kg/ms]	0.96049	0.8918
Thermal conductivity (W/m. °C)	0.6086	0.6105
Specific heat [kJ/kg. °C]	3.9954	4.18617

chemical composition of the open sea; however, the total amount of dissolved solids fluctuates depending on the conditions in the immediate area (El-Dessouky & Ettouney, 2002). This is because the diffusion time for salts, the time required for total mixing of all the seas and oceans, is substantially shorter than the time required for complete filling or replenishment. This is the reason why this is the case. A total salinity of 36,000 parts per million (ppm) is displayed in Table 1.1, illustrating saltwater's standard composition. Furthermore, seawater contains dissolved ions and a diverse range of suspended stuff such as clay, sand, viruses, bacteria, and colloidal particles (El-Dessouky & Ettouney, 2002) (Table 1.2).

1.3 WATER QUALITY PARAMETERS AFFECTING DESALINATION

The source water for desalination might consist of marine or brackish surface water and highly mineralized groundwater (Valle et al., 2014). This water inherently contains a substantial amount of inorganic ions, and the goal of treatment is to decrease or eliminate the concentration of these compounds (Omer, 1975). These compounds exist naturally and may be of concern if they are present in high concentrations after treatment. Surface water supplies and specific groundwater sources can get contaminated by harmful viruses, bacteria, parasites, and various chemical contaminants resulting from human activities (Yadav et al., 2022).

Sources of brackish or saltwater are distinct from those of freshwater. Particularly when combined with intense sunlight, saltwater drastically reduces the chances of survival of many microbial infections. But other diseases can actually thrive in saltwater, such as Vibrio cholerae (Kress et al., 2020; WHO, 2006). Additionally, a large number of marine algae have the potential to release harmful poisons into the environment (Yermiyahu et al., 2007). Desalination technology: health and environmental consequences delve deeply into these topics (Cotruvo, 2018). Boron (borate), sodium, iodide, bromide, and potassium are significant chemical components. Some of these may be difficult to remove (like boron), while others may be present in high enough concentrations to leave noticeable residues (Shah et al., 2019a). Although there is considerable variation in natural organic matter (NOM), several organic substances originating from anthropogenic and natural sources are particularly notable. Chemical, physical, and biological characteristics are the three main categories of water quality indicators (Cotruvo, 2018) (Table 1.3).

Figure 1.1 illustrates the components and processes of a water quality management system for effective monitoring and control.

According to Birnhack et al. (2010), among the quality parameters that should be taken into consideration for desalinated water are the following: pH, alkalinity, the water stability index as measured by calcium carbonate precipitation potential (CCPP), concentration ranges for Ca^{2+}, Cl, SO_4^{2-}, B, Na+, and Mg^{2+}, and electrical conductivity (EC).

The amount of dissolved solids is essential to establish whether or not it is suitable for use in irrigation, drinking, or industrial applications. The TDS number measures the water's overall salinity, with a higher value indicating a more salty taste (Tyagi et al., 2020). The dissolved solids in irrigation water are a highly essential criterion since their steady accumulation leads to the salinization of the soil, which in turn makes the agricultural area unproductive (Kumar et al., 2018).

Chemical oxygen demand (COD) is a method of determining how much organic matter is present in a given volume of water. This is because organic matter, derived from decomposing plant and animal waste, is the principal substance that dissolved oxygen in water oxidizes. Measuring the oxygen demand of both biodegradable and non-biodegradable pollutants is important to water quality assessment. COD levels in potable water typically fall between 1.2 and 2.6 mg/L (Vigil, 2023). Nevertheless, the COD level in fish intended for human consumption should not exceed 2 mg/L (Omer,

TABLE 1.3 Parameters of water quality

	TYPES OF WATER QUALITY PARAMETERS		
SR. NO.	CHEMICAL	BIOLOGICAL	PHYSICAL
1	pH	Algae	Taste and odor
2	Chloride	Viruses	Color
3	Chemical oxygen demand	Bacteria	Electrical conductivity
4	Biochemical oxygen demand	Protozoa	Turbidity
5	Chlorine residual		solids
6	Sulfate		Temperature
7	Copper and zinc		
8	Fluoride		
9	Toxic inorganic substances		
10	Nitrogen		
11	Alkalinity		
12	Dissolved oxygen		
13	Iron and manganese		
14	Acidity		
15	Hardness		
16	Toxic organic substances		
17	Radioactive substances		

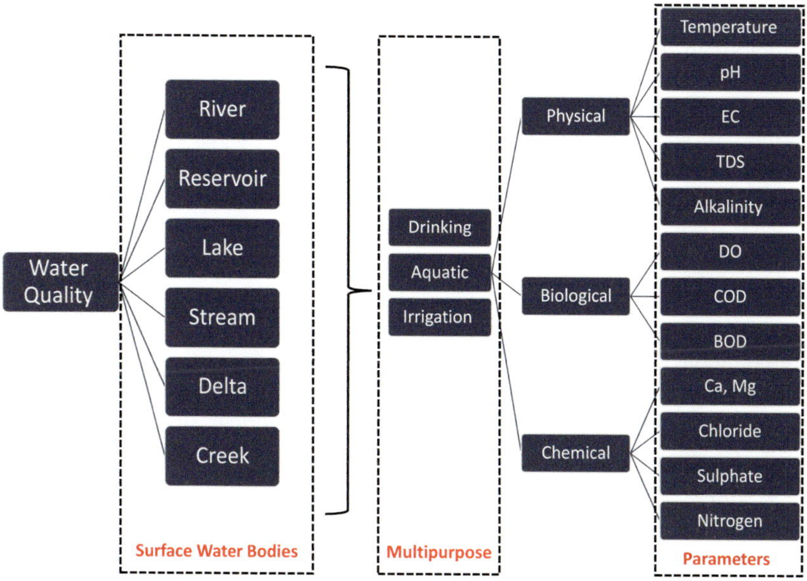

FIGURE 1.1 Surface water quality parameters.

1975). The concentration of COD in drinking water was found to be positively correlated with male and female liver cancer death rates, according to Alam. Effluent released into aquatic bodies requires an increasing amount of oxygen as the concentration of COD in the effluent rises. An increase in the rate of organic decomposition in the watershed due to human activities may be responsible for the raised COD level linked with pollution. These operations negatively impact water quality, including discharging sewage and agricultural runoff into the reservoir (Water, 2019).

Scale and Alkalinity water stability or general buffering capability is measured by its alkalinity. More alkalinity usually reduces the corrosion rate and lessens pH variations in dispersed water; however, too much alkalinity might result in excessive scale deposition where calcium may exist (Dasgupta et al., 2016). Alkalinity is a significant factor influencing pH because it is directly connected to the water's ability to buffer. The amounts of ions bicarbonate, carbonate, and hydroxide in water determine alkalinity. The water can better resist a pH change by releasing H+ and OH– ions if its alkalinity value is higher for a given pH value. Increased dissolved inorganic carbon (DIC) concentration of the carbonate species (CO^{2-}) corresponds to increased alkalinity at a given pH (Fewtrell, 2013). At higher pH values, an alkalinity that is too high, however, might accelerate the release of copper and lead metals. It is also recognized that should treatment be used, red water prevention can be achieved by keeping the system alkalinity high while taking into account a future pH shift. Scale deposition or corrosion can cause pH changes in non-stabilized finished water in the distribution system. The alkalinity content in distributed water should be kept above 1 milli-equivalent, or 60 mg/L ($CaCo_3$) (Graham, 1999).

"Total chlorine" (T-Cl) refers to the combined amount of free and combined chlorine. Accurate measurement of T-Cl levels is required. It is vital to choose the suitable remedies that should be given to water to sustain the residual amounts of free chlorine (Damodhar & Vikram Reddy, 2013; Shah et al., 2019b). Many local governments deliberately introduce chlorine into water supplies to effectively eliminate pathogenic microorganisms, such as viruses and bacteria, which have the potential to cause illness if consumed (Boyd, 2000). Natural water typically contains calcium from various rock types, industrial waste, and sewage. Mustapha states that oligotrophic waters typically have calcium levels below 10 mg/L, while eutrophic waters have values above 25 mg/L. Hard water may contribute to heart diseases, according to some research. Magnesium has a laxative effect and makes water taste bad when consumed in large quantities. Although water hardness is not a measure of pollution, it does show that the salinity is low because of the presence of ions such as $CaCO_3$, which stands for calcium and magnesium, and which are generally associated with bicarbonate and carbonate as well as sulfates, chlorides, and other anions of mineral acids. Because it causes scale to build on water heaters and cooking devices and uses more detergent when washing clothing, water with excess hardness is not ideal for human consumption. The increased hardness in the precipitation may be attributable to the greater magnesium and calcium ion concentration (Omer, 1975) (Table 1.4).

TABLE 1.4 Water parameter limits

				BIS (10500-2012)		
SR. NO.	*PARAMETERS*	*ID*	*UNITS*	*ACCEPTABLE LIMITS*	*PERMISSIBLE LIMITS*	*WHO*
1	Temperature	Temp.	0C	–	–	15–35
2	Alkalinity	Alk.	mg/L	200	600	—
3	Chloride	Cl	mg/L	250	1000	250
4	Sulfate	SO_4	mg/L	200	400	400
5	Total dissolved solids	TDS	mg/L	500	2000	1000
6	Magnesium	Mg	mg/L	30	100	—
7	Total hardness	TH	mg/L	200	600	—
8	Calcium	Ca	mg/L	75	200	—
9	Dissolved oxygen	DO	mg/L	4	6	
10	Potential of hydrogen	pH	–	6.5-8.5	No relaxation	6.5–8.5
11	Electrical conductivity	EC	µS/cm	300	800	300
12	Biochemical oxygen demand	BOD	mg/L	—	—	5
13	Nitrogen	NO_3	mg/L	45	—	—
14	Nitrogen as Ammonia	NH_3– N	mg/L	0.5	No relaxation	1.5
15	Chemical oxygen demand	COD	mg/L	—	—	10
16	Sodium	Na	mg/L	30	60	
17	Fluoride	F	mg/L	2		
18	Nitrogen as nitrite	NO_2-N	mg/L	—	—	—
19	Nitrogen as nitrate	NO_3-N	mg/L	45	No relaxation	10

A study conducted by Alam et al. (2007) showed that drinking water sources without conventional treatment but after purification had a biochemical oxygen demand (BOD) level of 2 mg/l or lower after five days at 20°C. According to Gupta et al. (2003), the amount of organic content increases when municipal and domestic garbage is discharged into untreated water bodies. Consequently, water's BOD rises due to bacteria in the water. BOD is a metric for assessing the pollution level in a body of water. According to Verma & Singh (2013), when BOD levels from decay exceed the water's local dissolved oxygen (DO) content, other species die due to a lack of oxygen.

The pH scale subjectively measures the amount of hydrogen ions in a solution and indicates the water's acidity or alkalinity level as well as the solution's basicity or acidity (Dohare et al., 2014). With a neutral pH of 7, the scale goes from 1 (acidic) to 14 (basic) (Chindwara et al., 2023). The pH range of 6.5 to 8.5 is considered standard for any use, including fish survival and potable water consumption [1; 4; 9]. Scale forms more quickly in water heating equipment, and chlorine's germicidal effectiveness is diminished at higher pH values. Chemical reactions, neutralization, sedimentation, coagulation, and corrosion control are only a few of the water treatment and delivery steps that affect the pH value (Garg & Joshi, 2015). The BOD needs to grow exponentially when acid rain or discharges alter the pH, which boosts bacterial decomposition, reduces DO, and ultimately increases the need for BOD (Graham, 1999).

1.4 BRACKISH GROUNDWATER COMPOSITION

TDS is a method for determining the concentration of all organic and inorganic substances in a given volume of water. Metals, minerals, and salts are all examples of such things. A certain quantity of TDS is present in all naturally occurring water. Using TDS, one can ascertain if a surface water or groundwater source is clean or contains salt (Ortiz et al., 2008). Compared to freshwater, brackish groundwater has a higher TDS concentration, while saltwater has a lower one. Various classification techniques are employed to categorize waters with differing TDS levels. Brackish groundwater can have a TDS content anywhere from 500 mg/L to 20,000 mg/L, with some classifications stating that 10,000 mg/L is the upper limit. However, TDS ratios in saltwater are frequently greater than 25,000 mg/L (Yvana, 2021) (Table 1.5).

The TDS and main ion components of brackish water, like seawater, can vary greatly depending on the geographical location and the depth of the well below the land surface (Kambanda, 2014). Local hydrologic, meteorological, and geologic factors all contribute to these differences. Deeper strata of the Earth's crust are usually used to recover groundwaters with higher TDS. The feasibility, treatment requirements, and associated costs of brackish groundwater can be better assessed with a thorough understanding of these variances (Krista et al., 2009).

Brackish water is present, with subsurface sources being the most significant. According to (Shanks et al., 1999), salty waters with TDS concentrations ranging from 1000 mg/L to 3000 mg/L cover more than half of the United States' geographical area.

Saltwater intrusion mostly happens in coastal locations by approaching close to discharging wells and lateral encroachment. The movement of saltwater toward a well can be caused by the generated gradients in places where groundwater is pumped from aquifers that are hydraulically connected to the sea. Additionally, groundwater withdrawals modify the patterns of groundwater flow and discharge to coastal habitats, which may influence the salinity and nutrient contents of the wetlands and coastal rivers (Kelly et al., 2001).

Applications that demand high-quality water, like irrigation and drinking water, cannot use brackish groundwater without prior cleaning. According to the US Environmental Protection Agency's recommendation, drinking water should have a TDS concentration of less than 500 mg/L (Inc, W. M.-G. I. 1988). This is because the public water supply, which fulfills most people's daily water requirements, should not include a high concentration of dissolved solids or large amounts of specific components. Dissolved solids concentration, relative solute amounts, and particular elements that can harm crops are the constraints on the water consumption by crops, which accounts for the vast majority of our world's freshwater consumption.

Groundwater levels are falling, and there is not enough safe, drinkable groundwater. This is a more significant challenge than water quantity problems alone regarding water stress. According to current estimates, 60% of the 300 km^3 of groundwater in the Ganges, Indus, and Brahmaputra River basin—widely recognized as one of the world's most extensive and productive groundwater systems—is considered unsuitable and dangerous (MacDonald et al., 2016).

Elevated levels of naturally occurring geogenic pollutants, such as arsenic (As) and fluoride (F), in groundwater, pose a significant worldwide health risk (Coomar & Mukherjee, 2021). Overall, they have exposed over 500 million individuals worldwide to possible significant risks to human health. The contamination of groundwater with arsenic has affected a population of 100 million people in the eastern regions of the United States, China, and South Asia (Haque et al., 2020).

Extended contact with water containing high levels of arsenic can lead to severe health issues, such as hyperkeratosis, melanosis, skin lesions, and cancer in various organs, which, in certain instances, can be life-threatening (Batabyal & Chakraborty, 2015). Contaminants can originate from aquifer geologic units, where trace elements like F and As are present in rocks. Previous studies have indicated a connection between As and the sediments in aquifers from nearby mountain chains, such as the Himalayas in South Asian countries and the Andes in South America. These sediments are deposited in alluvial plains and deltas through fluvial systems. On the other hand, F is believed to be associated with aquifers that are primarily found in crystalline or weathered bedrock systems. Due to surface weathering, these elements are mainly extracted from the host rocks, particularly granitites. Inland aquifers in several countries, including Argentina, Pakistan, Afghanistan, the Middle East, and India, contain very saline groundwater. Inland salinization can result via retaining ancient sea water as connate water, dissolving minerals through interactions with water and rocks, and infiltrating agricultural waste and sewage (Batabyal & Chakraborty, 2015).

The rapid expansion of industry and agriculture in the past century has significantly affected surface water and groundwater pollution with various chemical contaminants

TABLE 1.5 Brackish water limits

Physio-chemistry parameter	*RECOMMENDATIONS OF WHO*
Conductivity at °C (μScm^{-1})	-
pH	6.5–8.5
F^- (mgL^{-1})	1.5
Cl^- (mgL^{-1})	250
HCO_3^- (mgL^{-1})	-
NO_3^- (mgL^{-1})	50
SO_4^{2-} (mgL^{-1})	400
Na^+ (mgL^{-1})	250

(Mukherjee, 2019). Agricultural runoff is nonpoint source pollution that spreads out from several locations. Throughout the years, agricultural practices across the globe have extensively relied on agrochemicals to enhance crop production in their quest for increased productivity. Contaminants derived from agriculture that might infiltrate subsurface aquifers are frequently linked to fertilizers, including nitrate, phosphate-based, synthetic ones, and pesticides. Some agricultural pesticides, such as persistent organic pollutants and their by-products, can remain in groundwater for long. They tend to accumulate in organisms that consume the contaminated groundwater, increasing in concentration as they move up the food chain.

1.5 WATER QUALITY MANAGEMENT IN THE COASTAL ZONE

Seawater is the most valuable natural resource for coastal tourist areas that rely on water for recreational activities. The coastal zone interacts with economic, natural, and anthropogenic activities, making managing coastal zone resources difficult. Furthermore, disputes involving the utilization of coastal zone resources are frequently caused by these processes. The utilization of coastal waters is directly linked to most coastal ecosystems, including anthropogenic coastal zone systems. So, in managing the coastal zone, it is essential to consider the various seas (McCoy & Corbett, 2009).

The water quality management problem within the coastal zone can be divided into three categories worldwide. Identifying the qualitative qualities of the body of water is the first step in measuring the water body. It is essential to understand that, when it comes to such objects, there may be qualities shared by all of them, but there may also be characteristics unique to a particular aquatic object. The second issue is the formulation of water users' obligations toward one another. Third, water users must manage their processes to fulfill the standards set (Fewtrell, 2013).

Seawater physical-chemical parameters such as temperature, salinity, currents, waves, depth, turbidity, turbulence, residence times, wave exposure, upwelling, mixing characteristics, the spatial and temporal distribution of nutrients, oxygen, acidification, and turbidity are emphasized in the Marine Strategy Framework Directive. These factors are crucial to comprehending the anthropogenic-alterable dynamics of marine ecosystems (Yadav et al., 2022).

Location, topography, morphology, and lithology all have a role in hydrographic conditions, which are influenced by climate, ocean circulation, and large-scale forcings like tides (Valle et al., 2014). Particularly in coastal regions, river dynamics are affected by smaller-scale factors like land use and stresses caused by humans (Serrano-Tovar et al., 2019). Every coastal ocean observation system must have real-time data on the condition of the sea. It enables the assessment of long-term trends and the detection of anomalous events. To address the demands of different groups, the Gulf of Mexico Alliance (GOMA) suggested hundreds of sites in open Gulf of Mexico seas and estuaries around the coast to establish a Gulf Monitoring Network to collect data on water quality and other physical and biological variables. Details, including current speed, chlorophyll-a (Chla) concentrations, photosynthetically accessible radiation, dissolved oxygen, turbidity, pH, temperature, and salinity, are all part of this data set. Marine buoys, drifters, ships, and satellites are among the platforms suggested to gather this data at different frequencies, in addition to specialized sensors. However, substantial obstacles in terms of available resources (financial, human, and technological) make implementing such a monitoring network difficult (Gooch-Moore et al., 2011).

Conversely, recent satellite remote sensing offers frequent and synoptic data on several important water quality metrics. Over the past few decades, the ocean color community has advanced tremendously in algorithm development and sensor technologies. Coastal and estuarine water quality states may be evaluated regionally with enough accuracy using optical water quality data.

1.6 CONCLUSION

Water quality is crucial for drinking, irrigation, industrial processes, and ecological systems, among other uses, as the comprehensive study in this chapter highlights. It describes the many qualities of water temperature, color, taste, and smell—and how they affect its applicability for particular purposes. The chapter discusses the need to maintain water quality standards to protect the environment and human health. The chemical composition of brackish groundwater and seawater is also covered, emphasizing the difficulties with desalination and the necessity of efficient water quality control, especially in coastal areas. As it addresses the health issues related to water contamination, the study's conclusion emphasizes the need to manage water resources sustainably in order to ensure its availability and quality for present and future generations.

REFERENCES

Alam, M. J. B., Islam, M. R., Muyen, Z., Mamun, M., & Islam, S. (2007). Water quality parameters along rivers. *International Journal of Environmental Science and Technology*, *4*(1), 159–167. https://doi.org/10.1007/BF03325974

Amaaliya, N. K., & Kumar, S. P. (2013). *Carried out ground water quality status by water quality indexmethodatKanyakumari*.India.https://scholar.google.com/scholar?hl=en&as_sdt=0%2C5&q=Amaaliya+N+K+and+Kumar+S+P%2C+Carried+out+ground+water+quality+status+by+water+quality+index+method+at+K+anyakumari%2C+%28INDIA%29%2C+2%289%29%2C+76-82%2C+2013.+&btnG=

Batabyal, A., & Chakraborty, S. (2015). Hydrogeochemistry and water quality index in the assessment of groundwater quality for drinking uses. *Wiley Online Library*, *87*(7), 607–617. https://doi.org/10.2175/106143015X14212658613956

Birnhack, L., Penn, R., Oren, S., Lehmann, O., & Lahav, O. (2010). Pilot scale evaluation of a novel post-treatment process for desalinated water. *Desalination and Water Treatment*, *13*(1–3), 128–136. https://doi.org/10.5004/DWT.2010.984

Boyd, C. E. (2000). Overview of Water Pollution. Water Quality, and introduction. In *Water Quality: An Introduction*.

Chhaya, R. T., & Prajapati, S. K. (2020). Bioremediation of pharmaceuticals in water and wastewater. *Microbial Bioremediation & Biodegradation*, 425–446. https://doi.org/10.1007/978-981-15-1812-6_16

Chindwara, M. P., Jain, N., & Pahade, R. K. (2023). *Physico-Chemical Analysis of Drinking Water of.* https://scholar.google.com/scholar?hl=en&as_sdt=0%2C5&as_ylo=2013&as_yhi=2021&q=Physicochemical+Analysis+of+Pharmaceutical+Effluent+and+Surface+Water+of+River+Gorax+in+Minna%2C+Niger+State%2C+Nigeria%3A+Bulletin+of+Environment&btnG=%0Ahttps://scholar.goog

Coomar, P., & Mukherjee, A. (2021). Global geogenic groundwater pollution. *Global Groundwater*. https://www.sciencedirect.com/science/article/pii/B9780128181720000141

Cotruvo, J. (2018). Drinking Water Quality and Contaminants Guidebook. https://doi.org/10.1201/9781351110471/DRINKING-WATER-QUALITY-CONTAMINANTS-GUIDEBOOK-JOSEPH-COTRUVO

Damodhar, U., & Vikram Reddy, M. (2013). Impact of pharmaceutical industry treated effluents on the water quality of river Uppanar, South east coast of India: A case study. *Applied Water Science*, *3*(2), 501–514. https://doi.org/10.1007/s13201-013-0098-x

Darre, N. C., & Toor, G. S. (2018). Desalination of Water: A Review. *Current Pollution Reports*, *4*(2), 104–111. https://doi.org/10.1007/s40726-018-0085-9

Dasgupta, N., Ranjan, S., Chakraborty, A. R., Ramalingam, C., Shanker, R., & Kumar, A. (2016). Nanoagriculture and Water Quality Management. *Springer*, 1–42. https://doi.org/10.1007/978-3-319-39303-2_1

Dhok, R. P., Patil, A. S., & Ghole, V. S. (2011). Assessment of Ground Water Quality Status by Using Water Quality Index. *WaterR&D*, *1*(2 & 3), 72–78. https://search.proquest.com/openview/c82c132b5130b57e285dd2eaf59ac75e/1?pq-origsite=gscholar&cbl=2026591

Dohare, D., Deshpande, S., & Kotiya, A. (2014). Analysis of Ground Water Quality Parameters: A Review www.isca.me. *Research Journal of Engineering Sciences*, *3*(5), 2278–9472. http://www.isca.me/IJES/Archive/v3/i5/3.ISCA-RJEngS-2014-24.pdf

El-Dessouky, H. T., & Ettouney, H. Mohamed. (2002). *Fundamentals of Salt Water Desalination*. Elsevier. https://doi.org/10.1016/b978-0-444-50810-2.x5000-3

Fewtrell, L. (2013). Water Quality: Guidelines, Standards and Health: Assessment of Risk and Risk Management for Water-Related Infectious Disease. *Water Intelligence Online*, *12*, 1–431. https://doi.org/10.2166/9781780405889

Garg, M. C., & Joshi, H. (2015). Optimization and economic analysis for a small scale nanofiltration and reverse osmosis water desalination system. *Water Science and Technology: Water Supply*, *15*(5), 1027–1033. https://doi.org/10.2166/ws.2015.059

Ghenai, C., Kabakebji, D., Douba, I., & Yassin, A. (2021). Performance analysis and optimization of hybrid multi-effect distillation adsorption desalination system powered with solar thermal energy for high salinity sea water. *Energy*, *215*, 119212. https://doi.org/10.1016/j.energy.2020.119212

Gooch-Moore, J., Goodwin, K. D., Dorsey, C., Ellender, R. D., Mott, J. B., Ornelas, M., Sinigalliano, C., Vincent, B., Whiting, D., & Wolfe, S. H. (2011). New USEPA water quality criteria by 2012: GOMA concerns and recommendations. *Iwaponline.Com*, *9*(4). https://doi.org/10.2166/wh.2011.156

Graham, N. (1999). Guidelines for Drinking-Water Quality, 2nd edition, Addendum to Volume 1 – Recommendations, World Health Organisation, Geneva, 1998, 36 pages. *Urban Water*, *1*(2), 183. https://doi.org/10.1016/s1462-0758(00)00006-6

Gulamhussein, M. A., Shah, M., Yadav, K., Prajapati, M., & Shah, M. (2023). A comprehensive and systematic study on the techniques used for augmenting the performance of a solar still distillate yield. *Environmental Science and Pollution Research*, *30*(20), 57271–57278. https://doi.org/10.1007/s11356-023-26751-6

Gupta, A., Gupta, S., & Patel, R. (2003). A comparison of water quality indices for coastal water. *Journal of Environmental Science and Health, Part A*. https://www.tandfonline.com/doi/abs/10.1081/ESE-120024458

Haque, M. M., Niloy, M., Nayna, O. K., Fatema, K. J., Quraishi, S. B., Park, J.-H., Kim, K.-W., & Tareq, S. M. (2020). Variability of water quality and metal pollution index in the Ganges River, Bangladesh. *Environmental Science and Pollution Research*, *27*(34), 42582–42599. https://doi.org/10.1007/s11356-020-10060-3

Holland, H. D. (2005). Sea level, sediments and the composition of seawater. *American Journal of Science*, *305*(3), 220–239. https://doi.org/10.2475/ajs.305.3.220

Hu, C., Barnes, B. B., Murch, B., & Carlson, P. (2013). Satellite-based virtual buoy system to monitor coastal water quality. *Optical Engineering*, *53*(5), 051402. https://doi.org/10.1117/1.oe.53.5.051402

Inc, W. M.-G. I. (1988). United States Environmental Protection Agency. *Epa.Gov*. https://www3.epa.gov/npdes/pubs/owm0260.pdf

Jarin, M., Dou, Z., Gao, H., Chen, Y., & Xie, X. (2023). Salinity exchange between seawater/brackish water and domestic wastewater through electrodialysis for potable water. *Frontiers of Environmental Science and Engineering*, *17*(2), 1–10. https://doi.org/10.1007/s11783-023-1616-1

Kambanda, K. (2014). *Evaluation of desalination techniques for treating the brackish water of Olushandja sub-basin*. https://ubt.opus.hbz-nrw.de/opus45-ubtr/frontdoor/deliver/index/docId/664/file/Kambanda_Diss.pdf

Kelly, V., Hooper, R., Aulenbach, B., & Janet, M. (2001). Concentrations and annual fluxes for selected water-quality constituents from the USGS National Stream Quality Accounting Network (NASQAN), 1996–2000. *Pubs.Usgs.Gov*. https://pubs.usgs.gov/wri/wri014255/org.htm

Khawaji, A. D., Kutubkhanah, I. K., & Wie, J. M. (2008). Advances in seawater desalination technologies. *Desalination*, *221*(1–3), 47–69. https://doi.org/10.1016/j.desal.2007.01.067

Koroneos, C., & Roumbas, G. (2012). Geothermal waters heat integration for the desalination of sea water. *Desalination and Water Treatment*, *37*(1–3), 69–76. https://doi.org/10.1080/19443994.2012.661256

Kress, N., Gertner, Y., & Shoham-Frider, E. (2020). Seawater quality at the brine discharge site from two mega size seawater reverse osmosis desalination plants in Israel (Eastern Mediterranean). *Water Research, 171*. https://doi.org/10.1016/j.watres.2019.115402

Krista, D., Kulinkina, A., & Plummer, J. D. (2009). *The Design of a Desalination Pretreatment System for Brackish Groundwater a Major Qualifying Project Report.*

Kucera, J. (2014). *Desalination: Water from Water* (pp. 1–624). https://doi.org/10.1002/9781118904855

Kumar, R., Ahmed, M., Bhadrachari, G., & Thomas, J. P. (2018). Desalination for agriculture: Water quality and plant chemistry, technologies and challenges. *Water Supply, 18*(5), 1505–1517. https://doi.org/10.2166/ws.2017.229

Ludwig, H. (2022). *Reverse Osmosis Seawater Desalination Volume 1* (Vol. *1*). https://doi.org/10.1007/978-3-030-81931-6

MacDonald, A., Bonsor, H., Ahmed, K., Burgess, W. et al. (2016). Groundwater quality and depletion in the Indo-Gangetic Basin mapped from in situ observations. *Nature.Com*. https://www.nature.com/articles/ngeo2791

McCoy, C. A., & Corbett, D. R. (2009). Review of submarine groundwater discharge (SGD) in coastal zones of the Southeast and Gulf Coast regions of the United States with management implications. *Journal of Environmental Management, 90*(1), 644–651. https://doi.org/10.1016/j.jenvman.2008.03.002

Miller, W. W., Johnson, D. W., Karam, S. L., Walker, R. F., & Weisberg, P. J. (2010). A synthesis of Sierran forest biomass management studies and potential effects on water quality. *Forests, 1*(3), 131–153. https://doi.org/10.3390/f1030131

Mukherjee, A. (2019). Changing the groundwater landscape of India: Implications to drinking water, food security, socio-economy and public health. *Proceedings of the National Academy of Sciences*. https://doi.org/10.16943/ptinsa/2019/49708

Omer, N. H. (1975). Water Quality Parameters. *ASTM Special Technical Publication, STP 573*, 1–18. https://doi.org/10.1201/9781439800539.axb

Ortiz, J. M., Expósito, E., Gallud, F., García-García, V., Montiel, V., & Aldaz, V. A. (2008). Desalination of underground brackish waters using an electrodialysis system powered directly by photovoltaic energy. *Solar Energy Materials and Solar Cells, 92*(12), 1677–1688. https://doi.org/10.1016/j.solmat.2008.07.020

Panagopoulos, A. (2021). Water-energy nexus: desalination technologies and renewable energy sources. *Environmental Science and Pollution Research, 28*(17), 21009–21022. https://doi.org/10.1007/s11356-021-13332-8

Prajapati, M., Shah, M., Soni, B., Parikh, S., Sircar, A., Balchandani, S., Thakore, S., & Tala, M. (2021). Geothermal-solar integrated groundwater desalination system: Current status and future perspective. In *Groundwater for Sustainable Development* (Vol. *12*, p. 100506). Elsevier B.V. https://doi.org/10.1016/j.gsd.2020.100506

Rajkamal, R., Kumar, M., Raj, M., Rajesh, M., Kiruthiga, J., & Bazroy, J. (2016). Assessment of water quality standards in the villages of Kanchipuram district, Tamil Nadu, India. *International Journal of Community Medicine and Public Health, 3*(11), 3179–3183. https://doi.org/10.18203/2394-6040.ijcmph20163932

Sarda, P., & Sadgir, P. (2015). Assessment of Multi Parameters of Water Quality in Surface Water Bodies-A Review. ISSN: 2321-9653 https://www.researchgate.net/publication/282356694. *International Journal for Research in Applied Science & Engineering Technology (IJRASET), 3*(October), 331–336.

Serrano-Tovar, T., Peñate Suárez, B., Musicki, A., de la Fuente Bencomo, J. A., Cabello, V., & Giampietro, M. (2019). Structuring an integrated water-energy-food nexus assessment of a local wind energy desalination system for irrigation. *Science of the Total Environment, 689*, 945–957. https://doi.org/10.1016/j.scitotenv.2019.06.422

Shaffer, D. L., Yip, N. Y., Gilron, J., & Elimelech, M. (2012). Seawater desalination for agriculture by integrated forward and reverse osmosis: Improved product water quality for potentially less energy. *Journal of Membrane Science, 415–416*, 1–8. https://doi.org/10.1016/j.memsci.2012.05.016

Shah, M., Sircar, A., Varsada, R., Vaishnani, S., Savaliya, U., Faldu, M., Vaidya, D., & Bhattacharya, P. (2019a). Assessment of geothermal water quality for industrial and irrigation purposes in the Unai geothermal field, Gujarat, India. *Groundwater for Sustainable Development, 8*, 59–68. https://doi.org/10.1016/j.gsd.2018.08.006

Shah, M., Sircar, A., Varsada, R., Vaishnani, S., Savaliya, U., Faldu, M., Vaidya, D., & Bhattacharya, P. (2019b). Assessment of geothermal water quality for industrial and irrigation purposes in the Unai geothermal field, Gujarat, India. *Groundwater for Sustainable Development, 8*, 59–68. https://doi.org/10.1016/j.gsd.2018.08.006

Shanks, G. et al. (1999). Understanding data quality: Social and cultural aspects. *Proceedings of the 10th Australasian Conference on Information Systems, 1999.* https://www.researchgate.net/profile/Brian-Corbitt/publication/228607385_Understanding_data_quality_Social_and_cultural_aspects/links/00b4952b0f62468150000000/Understanding-data-quality-Social-and-cultural-aspects.pdf

Singh, A., Panda, S. N., Uzokwe, V. N. E., & Krause, P. (2019). An assessment of groundwater recharge estimation techniques for sustainable resource management. *Groundwater for Sustainable Development, 9*(April), 100218. https://doi.org/10.1016/j.gsd.2019.100218

Tyagi, S., Sharma, B., Singh, P., & Dobhal, R. (2020). Water Quality Assessment in Terms of Water Quality Index. *American Journal of Water Resources, 1*(3), 34–38. https://doi.org/10.12691/ajwr-1-3-3

Valle, R. F., Varandas, S. G. P., Sanches Fernandes, L. F., & Pacheco, F. A. L. (2014). Groundwater quality in rural watersheds with environmental land use conflicts. *Science of the Total Environment, 493*, 812–827. https://doi.org/10.1016/j.scitotenv.2014.06.068

Verma, A. K., & Singh, T. N. (2013). Prediction of water quality from simple field parameters. *Environmental Earth Sciences, 69*(3), 821–829. https://doi.org/10.1007/S12665-012-1967-6

Vigil, K. M. (2023). *Clean Water, 2nd ed: An Introduction to Water Quality and Water Pollution Control.* https://doi.org/10.1353/book83960

Water, N. (2019). *Drinking water quality.* https://niopa.qub.ac.uk/bitstream/NIOPA/9434/4/2017niwaterdrinkingwaterqualityannualreport.pdf

WHO. (2006). Guidelines for Drinking-water Quality FIRST ADDENDUM TO THIRD EDITION Volume 1 Recommendations, WHO Library Cataloguing-in-Publication Data. In *World Health Organization: Geneva.*

WHO. (2011). Safe Drinking-water from Desalination. *World Health Organization, WHO/HSE/WS*(1), 34. http://apps.who.int/iris/bitstream/10665/70621/1/WHO_HSE_WSH_11.03_eng.pdf

WMO, & Secretariat, WMO. (1988). *Manual on water-quality monitoring: planning and implementation of sampling and field testing.* 197. https://digitallibrary.un.org/record/29682

Yadav, S., Negm, A. M., Narayan, R., & Editors, Y. (2022). *Earth and Environmental Sciences Library Water Quality, Assessment and Management in India.*

Yermiyahu, U., Tal, A., Ben-Gal, A., Bar-Tal, A., Tarchitzky, J., & Lahav, O. (2007). Environmental science: Rethinking desalinated water quality and agriculture. *Science, 318*(5852), 920–921. https://doi.org/10.1126/science.1146339

Yvana, D. (2021). *Desalination of brackish groundwater to Publisher.* Elsevier.

A Critical Overview of Desalination Technologies and Sustainable Energy

2

2.1 INTRODUCTION

The sources of energy and water are considered essential components that are required to sustain human survival (Tzen & Morris, 2003). Because water and energy seem very important to humans, the energy-water nexus has received much attention nowadays. A growing global population will put enormous demand on the world's existing system for delivering energy and potable water (Kalogirou, 2005). Since water comprises two-thirds of the Earth, its accessibility is a significant factor in environmental sustainability. Nevertheless, just 1% of the world's freshwater sources are fit for human use (Abdelkareem et al., 2018).

The lack of available water is one of the most prominent global challenges we are confronted with in the twenty-first century. It has been stated by Darre and Toor (2018); Mohammadzadeh Bina et al. (2018) that the existing water resources are insufficient and continuously disappearing due to rapid industrialization, population growth, excessive exploitation, and climate change. The United Nations (UN) estimates that by the year 2050, about seven billion people in 60 countries will suffer from acute water scarcity (Ihsanullah et al., 2021). According to Mavukkandy et al. (2019) and Panagopoulos (2021), seawater desalination is considered one of the most viable and oldest techniques that may meet the growing need for high-quality water. The International Desalination Association (IDA) has published research indicating that 19,744 desalination units worldwide can desalinate around 99.7 million m^3 of water per day. As per Voutchkov and Kaiser (2019), these plants are located in 150 nations worldwide and currently serve the needs of over 300 million people. According to Darre and Toor (2018) and Petersen et al. (2018), the supply of desalinated water is expected to increase by a factor of three

DOI: 10.1201/9781003530893-2

by the year 2050. Based on Jones et al. (2019) and van Wyk et al. (2020), the global brine output is estimated to be 141.5 million m^3 daily, mostly discharged into ocean ecosystems.

Seawater desalination provides numerous benefits to humanity, including the availability of a consistent and high-quality potable water source and the conservation of natural water resources (Ameri et al., 2009). Additionally, it offers various environmental, socio-economic, and health advantages. Although desalination offers various advantages, there is ongoing discussion on the potential adverse environmental impacts (EIs) associated with this process (Miller et al., 2015).

Water is usually solely considered in its liquid form but may also exist in solid ice and gaseous vapor (Almtairi et al., 2021; Lee et al., 2011). It is estimated that water covers 71% of the surface of the Earth, most of which is found in seas and other major water bodies, with another 1.6% found in aquifers and 0.001% found in the air as vapor, clouds (made up of liquid and solid water molecules suspended in the atmosphere), and precipitation (Kucera, 2014). The world's saltwater oceans account for 97.0% of the planet's surface water, the polar ice caps and glaciers for 2.4%, and rivers and lakes on land for the remaining 0.6%. All of these variations of water eventually make their way to the ocean via the never-ending water cycle of evaporation, precipitation, transpiration, and runoff (Al-Ghamdi et al., 2014).

On the other hand, the number of desalination plants has only increased by a ratio of one and a half over 20 years (Gooch-Moore et al., 2011). This is even though the output capacity of desalination plants has increased by six. This may be because desalination has positive and negative effects despite technology's many advantages (such as making freshwater readily available in arid and coastal regions) (Kress, 2019). The desalination process has a few significant downsides. First, it uses a lot of energy. Second, it produces brine, which is a highly salty by-product. Fossil fuels typically supply the energy needed to run desalination plants (Elimelech & Phillip, 2011).

Each desalination system necessitates distinct forms of energy to facilitate its operation (Pouyfaucon & García-Rodríguez, 2018). Due to the negative EIs associated with traditional energy sources, researchers are increasingly considering using renewable energy sources (RES) to power desalination plants (Brower et al., 2014). Therefore, researchers are exploring different desalination designs that utilize RES as a possible alternative. According to Abdelkareem et al. (2018), the selection of suitable RES for desalination is determined by several factors. These factors include the economic feasibility of producing freshwater, the characteristics and pressure of feedwater, the location of the desalination plant, and its size. The following sections outline different categories of RES that can be incorporated into desalination systems (Chekifi et al., 2024; Nelson, 2011).

The chapter critically examines desalination technologies as a response to global water scarcity, highlighting the energy-water nexus and the increasing demand for freshwater due to population growth and industrialization. It explores the evolution of desalination from traditional thermal distillation methods like multi-stage flash (MSF) and multi-effect distillation (MED) to modern membrane processes such as reverse osmosis (RO). The chapter emphasizes the significant role of RES in powering desalination plants to mitigate EIs and promote sustainability. With a detailed analysis of various RES integration with desalination systems, the study underscores the potential of solar, wind, and geothermal energy in addressing water scarcity while reducing reliance on fossil fuels.

Section 2.2 provides a comprehensive overview of desalination technologies and their integration with sustainable energy sources. It begins by highlighting the critical importance of water and energy for human survival and the increasing pressures on freshwater and energy delivery systems due to population growth. The section discusses water scarcity as a significant global concern and the role of desalination as a viable solution to meet the high-quality water demand. The section further explores various desalination technologies, including thermal distillation and membrane separation, and their global distribution and capacity. Section 2.3 delves into the energy-intensive nature of desalination processes, which convert seawater into freshwater while discharging concentrated saltwater back into the ocean. The focus is on research efforts to improve energy-efficient desalination devices to address the issue of high energy consumption. The section also discusses the direct correlation between the energy and yield efficiency of desalination systems and the specific engineering designed to maximize user cost efficiency. Section 2.4 discusses the integration of sustainable energy sources with desalination technologies. It highlights the growing significance of energy in society for maintaining a high standard of living and operating other economic components. The section emphasizes the challenges posed by the increasing use of fossil fuels, such as ozone layer depletion, climate change, environmental issues, and health hazards. The section also explores how RES can produce energy without air pollutants or greenhouse gases and the benefits of decentralized renewable energy harvesting for rural and small-scale communities. Section 2.5 discusses case studies and best practices in the context of water desalination. It highlights the undervaluation of water supplies in many regions and the challenges water scarcity poses to sustainable development. The section emphasizes the importance of identifying and developing new water resources as the global population is projected to increase significantly. It presents case studies from various regions, including the Middle East, Australia, Singapore, and the United States, showcasing different approaches to addressing water scarcity through desalination technologies. Section 2.6 discusses the future directions and research opportunities in desalination systems. It emphasizes the importance of evaluating the sustainability of these systems through environmental, economic, social, and technological aspects. The section highlights the need for technical feasibility and the role of technology in reducing environmental emissions and energy consumption. It also touches on the economic impact of desalination, considering factors like water source quality, energy source type, labor costs, and geographical location. It suggests that RES integrated with advanced desalination technologies could provide a promising solution to global water crisis issues (Figure 2.1).

Figure 2.1 Pie charts showing how water is distributed around the world. Ninety-seven percent of the water in the largest part is saline. Three percent is freshwater, which is divided into surface water (1%), groundwater (30%), and icecaps and glaciers (68%). Lakes make up 87% of fresh surface water, followed by wetlands at 11% and rivers at 2%.

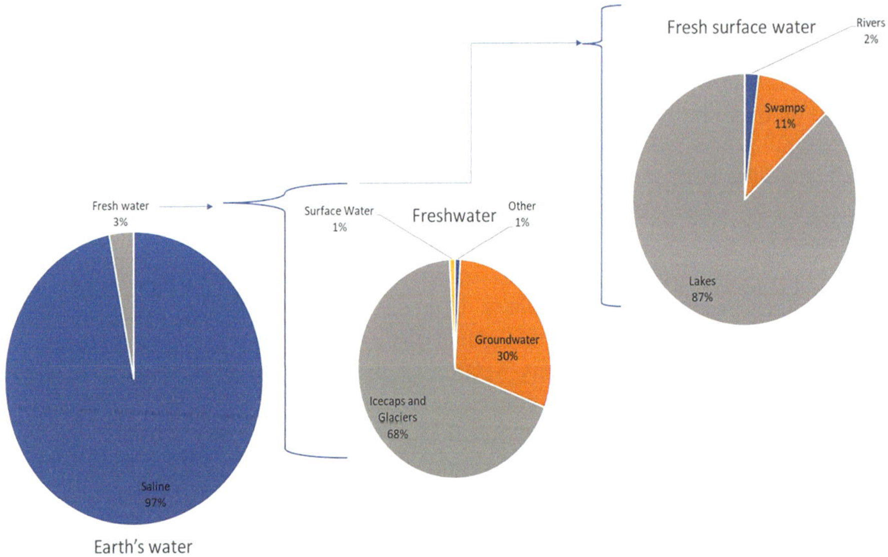

FIGURE 2.1 Distribution of Earth's water.

2.2 TYPES OF DESALINATION TECHNOLOGIES

The technology employed in commercial desalination can be categorized into two main groups (Bundschuh et al., 2021): technologies for desalination based on thermal processes (Gude et al., 2010), which include MED (Alhaj et al., 2017) and MSF distillation (Darwish, 1991), and membrane-based desalination (Mbarga et al., 2014), which includes RO separation. Both membrane and thermal technologies are incorporated into the growth of certain hybrid plants (Esmaeilion, 2020). Capacitive deionization (Huang et al., 2017), forward osmosis, electrodialysis reversal/electrodialysis (EDR/ED) (Aende et al., 2020), membrane distillation (Alklaibi & Lior, 2005), humidification-dehumidification (Lawal et al., 2018), and freezing nano/ultrafiltration/ionic, solar stills (Vishwanath Kumar et al., 2015) are some of the other technologies that are currently being utilized or are still in the early stages of development (Figure 2.2).

Figure 2.2 Flowchart depicting desalination techniques, classified into conventional and alternative approaches. Commonly utilized techniques include thermal or phase change procedures, including MED, MSF, and thermal vapor compression, as well as membrane processes like RO and ED. Alternative methods include membrane distillation, freezing, and ion exchange procedures.

Based on the statistics in Figure 2.6, the North Africa and Middle East region possesses the largest quantity of desalination plants, with 4826.0 plants (Andersen & Bank, 2012). This represents 47.5% of the total desalination capacity worldwide. The regions of Eastern Europe and Central Asia collectively possess a total of 566.0 plant species,

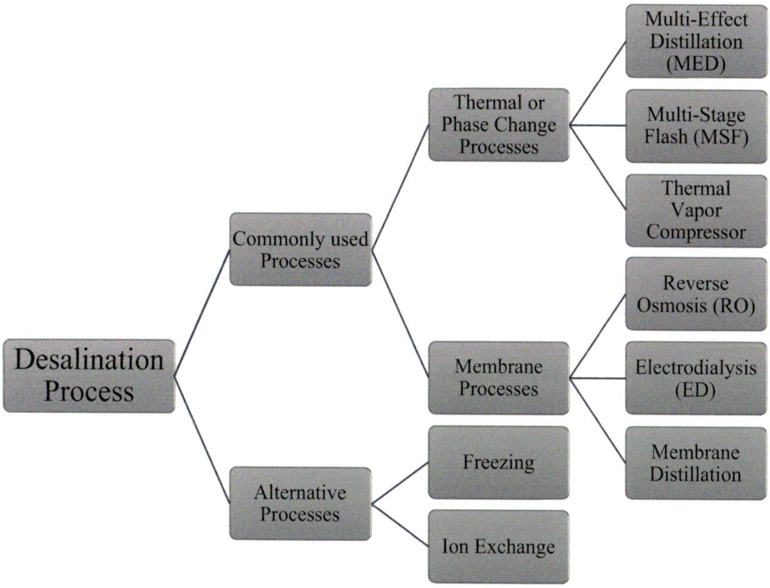

FIGURE 2.2 Desalination processes (Bhagwati et al., 2023).

accounting for 2.4% of the global plant diversity. Sub-Saharan Africa has 303.0 plants, which is around 1.9% of the global capacity. Furthermore, nations like the United Arab Emirates, Israel, and the Kingdom of Saudi Arabia depend on desalinated water to meet more than 50% of their water needs. Desalinated water is used for various reasons, including industrial and home consumption, energy production, agriculture, defense, and other applications (Gude, 2016).

Around 70% of worldwide desalination facilities use RO as their primary technology. MED (7%) and MSF (21%) based on thermal energy are the most widely used technologies (Prajapati et al., 2022). The components of a desalination system consist of NF (nanofiltration) and ED/EDR (Global water intelligence, 2011). The compactness and energy efficiency of membrane-based technologies provide them an advantage over thermal-based technologies. Additionally, membrane-based technologies offer lower salinity restrictions for the feed water than thermal-based technologies (Panagopoulos, 2021). As shown in Figure 2.3, a significant portion of the total production capacity is attributed to RO, followed by MSF at 18–21%, MED at 5–7%, ED at 3%, and various desalination methods at 4% (Prajapati et al., 2022).

Figure 2.3 Pie chart illustrating percentage distribution of production capacity of desalination process: RO at 65% in green color, MSF at 21% in blue color, MED at 7% in yellow color, ED at 3% in dark green color, and OTHER at 4% in dark blue color.

Research on membrane and thermal desalination systems has progressed substantially in the past several years. The use of membrane distillation, adsorption desalination, pervaporation, humidification and dehumidification, forward osmosis, and semi-batch RO processes are some of the newer technologies that are gaining popularity

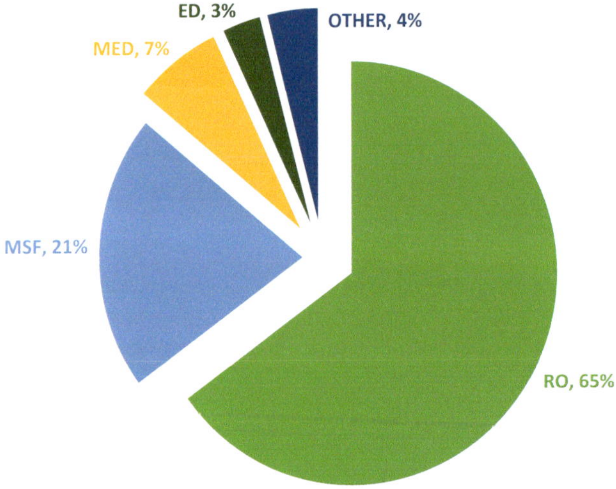

FIGURE 2.3 Desalination facilities worldwide use.

(Subramani & Jacangelo, 2015). Nanocomposite, nanotube, aquaporin, and graphene-based membranes are examples of recent advances in membrane technology.

2.2.1 Multi-stage Flash Desalination

MSF is a thermal distillation method that employs the principles of condensation and evaporation in order to separate water (Darwish, 1991). The processes of evaporation and condensation are interconnected at different phases, allowing for the recovery and reutilization of the latent heat of evaporation (G. et al., 2011). This is achieved by subjecting the entering water to preheating, as shown in Figure 2.4. The top brine temperature is the most significant temperature that incoming feed water can reach; this is achieved by the brine heater, which uses saturated steam condensed from the cold end of a steam cycle power plant or some other heat source. This process is illustrated in Figure 2.4. After that, the heated saltwater is pumped into the initial stage of evaporation, which has a lower pressure than the previous stage. The sudden entrance of hot water into the chamber is accompanied by a decrease in pressure, which causes the water to boil rapidly, almost causing it to flash into steam. As the flashing process continues, the vapor produced is condensed on the heat exchanger tubes running through the upper portion of each step (Nafey et al., 2006). The tubes undergo cooling with the introduction of feed water, which then flows toward the brine heater. The recovery of a portion of the thermal energy expended during the initial evaporation stage can be achieved by preheating the water. An MSF unit operates at a gradually reduced pressure at each consecutive stage, intending to optimize water and energy recovery (Lian-Ying et al., 2012).

In conjunction with a mechanical vacuum pump, a high-pressure steam-powered steam ejector can effectively sustain the vacuum at the intended level. MSF units are

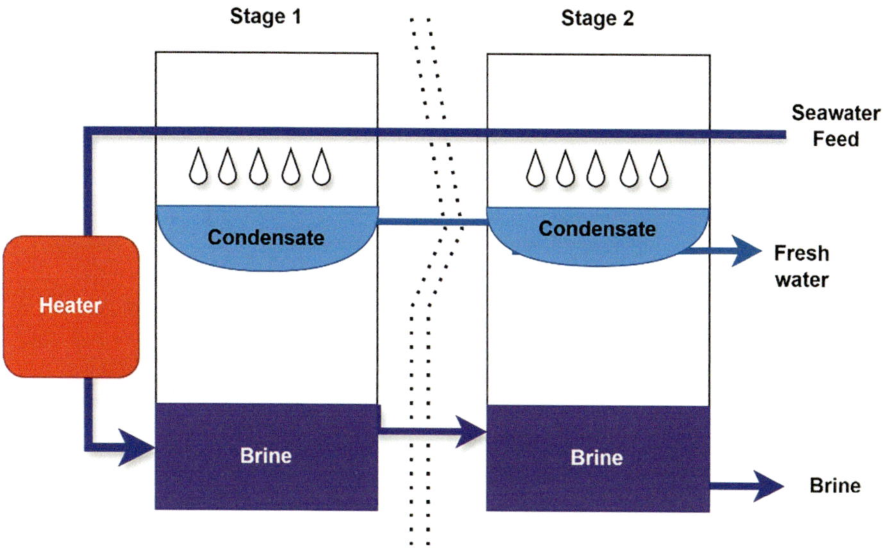

FIGURE 2.4 Multi-stage flash desalination.

widely employed in Saudi Arabia, the United Arab Emirates, the Middle East, and Kuwait (Al-Dousari et al., 2012). These units are responsible for 34% of the total desalination of seawater that occurs worldwide. The boiling of liquids in large quantities is essential to MSF systems. Concerns regarding the creation of scale on heat transfer tubes have been resolved (Belessiotis et al., 2016).

Figure 2.4 illustrates the MSF desalination process, which includes different parts such as seawater feed inlet, heater, freshwater outlet, and brine outlet.

2.2.2 Multi-effect Distillation

MED is today's earliest industrial desalination technique (Engelien & Skogestad, 2005). The MED includes several evaporators, called effects, that are subjected to a progressive decrease in temperature and pressure from the beginning to the end, with temperatures typically ranging from 60°C to 90°C (Bamufleh ct al., 2017; Cui et al., 2019). The lower pressure in each effect makes water evaporating at lower temperatures possible. This is because the boiling point of water drops as pressure increases. Each effect is made up of bundles of evaporator tubes that are sprayed with seawater. In the initial effect, the tubes are used to heat steam or hot water, and in each subsequent effect, the seawater receives energy from the tubes, which results in a partial evaporation of the water. The final effect of utilizing a heat exchanger to condense steam is producing clean distillate water. Product water or freshwater is being put into a storage tank, and brine is being pushed back into the ocean (Hatzikioseyian et al., 2003). The average amount of thermal energy required by MED plants that produce water in the range of 5000 to 50,000 m^3 per day is

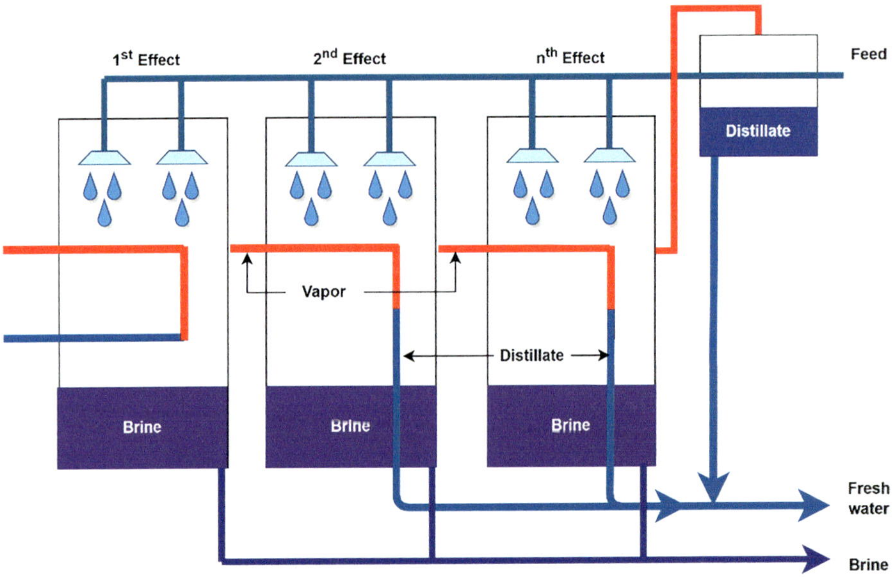

FIGURE 2.5 Multi-effect distillation.

between 12.2 and 19.1 kWh/m^3 of electrical energy. In addition, pumps are consumed for an extra 2–2.5 kWh/m^3 of electrical energy (Prajapati et al., 2022). The energy that MED uses directly affects the feed and vapor flow design. The three basic variations are forward feed, reverse feed, and parallel feed (Hussain et al., 2017).

Figure 2.5 illustrates the MED process, which includes different parts such as number of effects, feed inlet, freshwater outlet, and brine outlet.

Figure 2.6 illustrates desalination plants and global capacity. The North Africa and Middle East region has the largest number of desalination plants, accounting for 47.5% of the global capacity. Eastern Europe and Central Asia have 566.0 plants, while Sub-Saharan Africa has 303 plants. Countries like the UAE, Israel, and Saudi Arabia rely on desalinated water for over 50% of their water needs, used in industrial, energy, agriculture, and defense.

2.2.3 Reverse Osmosis

RO is a novel method that achieved commercial viability for water desalination in the early 1970s (Edzwald & Haarhoff, 2011). Membrane separation technology, RO, is water extraction from a pressurized saline solution by its passage across a membrane. This process effectively separates the water from the dissolved material without heating or changing its phase. The primary energy need is for the pressurization of the feed water. Alternatively, it can be defined as applying pressure greater than the osmotic pressure to move a solvent from an area with a high solute concentration to an area with a low solute concentration by passing it across a membrane (Ludwig, 2022).

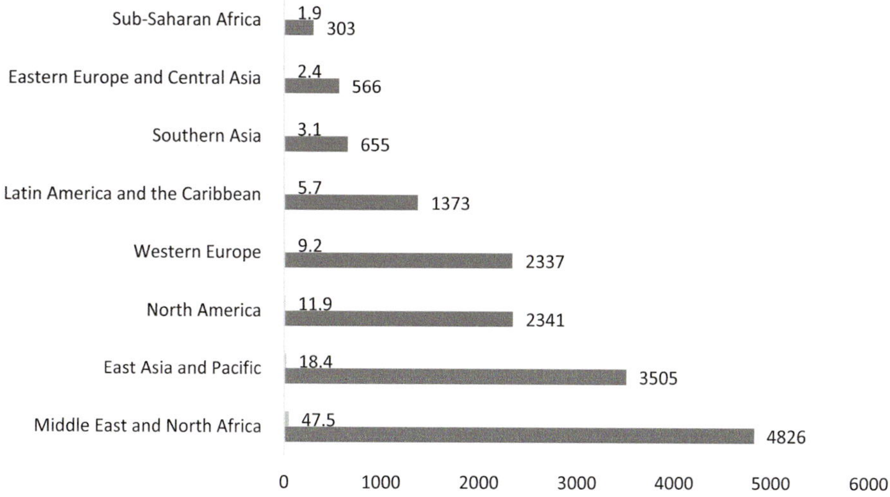

	Middle East and North Africa	East Asia and Pacific	North America	Western Europe	Latin America and the Caribbean	Southern Asia	Eastern Europe and Central Asia	Sub-Saharan Africa
Global Desalination Capacity (%)	47.5	18.4	11.9	9.2	5.7	3.1	2.4	1.9
Plants	4826	3505	2341	2337	1373	655	566	303

Global Desalination Capacity (%) Plants

FIGURE 2.6 Desalination plants and global capacity.

Consequently, water passes in the opposite direction of the inherent flow across the membrane, leading to an increased concentration of dissolved salts (Feria-Díaz et al., 2021). A conventional large-scale RO plant comprises five fundamental components: a saline water supply system, a feed water pretreatment system, a high-pressure pump, RO modules employing membrane separation, and a post-treatment system. Saudi Arabia and other Middle Eastern countries have recently become home to the world's largest RO desalination plants. The Jeddah facility can produce 15 MGD daily, but the Al Jubail and Yanbu RO plants can produce 24 MGD and 33.8 MGD, respectively (*Acwa Power Inaugurates Jubail 3a Iwp (Jazlah Water Desalination Company), the First Large-Scale Integrated Water Desalination—Solar PV Project in Saudi Arabia*, 2023; Baig & Al Kutbi, 1999).

2.2.4 Electrodialysis

ED employs two distinct types of membranes. The cation membrane selectively permits the passage of cations (+ve charged ions), while the anion membrane selectively permits the passage of anions (-ve charged ions) (Al-Amshawee et al., 2020; AlMadani, 2003).

At the same time as the exchange membranes are submerged in salty water in parallel, an electric current is passed across the liquid. There will be a strong attraction between positively charged ions and the negatively charged electrode, referred to as the cathode. On the other hand, negatively charged ions will be drawn to the positively charged electrode, referred to as the anode (Alklaibi & Lior, 2005).

Consequently, the water that flows through the membranes is divided into two separate streams. One side is entirely freshwater, whereas the other is a highly concentrated salt solution. ED is particularly advantageous for desalting brackish water due to its direct correlation between energy consumption and the amount of salts to be removed. RO and electrodialysis are the two predominant membrane processes. While ED employs electric current, RO utilizes hydraulic pressure to induce salt separation (Elimelech & Phillip, 2011; García-Rodríguez, 2003).

2.3 ENERGY REQUIREMENTS IN DESALINATION TECHNOLOGY

Desalination is a very energy-consuming procedure that utilizes energy to convert seawater into freshwater and discharges the concentrated saltwater back into the ocean (Shahzad et al., 2017). Loganathan et al. (2019) note that as desalination activities and huge facilities spread around the globe, worries over the process's possible detrimental effects on the environment are on the rise. Researchers and developers are investigating improvements in energy-efficient desalination devices to tackle the issue of high energy usage (Goh et al., 2017).

Most research efforts primarily concentrate on the energy and yield efficiency of desalination systems, while insufficient attention is given to the requirements of the industrial sector. Within the industry, all desalination systems are specifically engineered to maximize the efficiency of the whole cost to the user rather than focusing solely on energy consumption (Al-Karaghouli & Kazmerski, 2013). Despite extensive study in this field, the energy consumption of water desalination methods has remained mostly unchanged over the previous decade (Table 2.1).

There are many ways thermal desalination systems can be more energy efficient. Evaporation and condensation work in combination to allow the reuse and recycling of latent heat over many stages, which is the primary cause of this. Like the MSF process, the MED procedure functions through a multi-stage operation. Extremely few research have focused on the exergy evaluation of the MED process. A computer program was developed to evaluate the exergy efficiency of a solar-powered multi-effect distillation (MED) desalination plant (García-Rodríguez, 2003; Garcia-Rodriguez et al., 2002). According to the study, using energy recovery devices can increase the desalination process' exergy efficiency from 14.3% to 25.7%. Another study by Mabrouk and Fath (2015), Mezher et al. (2011), and Nafey (1988) used exergetic analysis to evaluate a MED desalination facility with a 15,000 m^3/day capacity. In this case, the research

TABLE 2.1 Energy consumed by various desalination processes (Abdelkareem et al., 2018)

	DESALINATION PROCESSES						
CHARACTERISTICS	*MSF*	*MED*	*MVC*	*TVC*	*SWRO*	*BWRO*	*ED*
Capacity of unit (m^3/day)	50,000–70,000	5000–15,000	100–3000	10,000–30,000	maximum to 128,000	maximum to 98,000	2–145,000
Thermal energy (kWh/m^3)	15.83–23.50	12.2–19.10	-	14.50	-	-	-
Electrical energy in (kWh/m^3)	2.5–5	2–2.5	7–12	1.6–1.8	4 to 6 with energy recovery	1.5–2.5	2.64–5.5
Total energy usage in (kWh/m^3)	19.58–27.25	14.45–21.35	7–12	16.20	4 to 6	1.50–2.5	2.64–5.5
Water product value in (ppm)	Almost to ten	Almost to ten	Almost to ten	Almost to ten	400–500	200–500	150–500

shows that the amount of effects strongly influences the exergy efficiency of the desalination process.

The exergy efficiency significantly rose, rising from 3.8% to 8.4%, when the number of impacts was augmented from 1 to 6. Unlike MSF procedures, MED processes operate at significantly lower temperature ranges (Panagopoulos, 2021). The temperature differential between the top brine temperature (TBT) in the first stage and the lowest brine temperature (LBT) in the last stage is reduced as a result of this, which reduces the number of stages utilized for the recovery of energy and product. Increasing the temperature differential between the TBT and the LBT is usual practice from 70°C–40°C to 70°C–10°C. This is done to maximize the potential number of evaporation phases and impacts for thermal processes such as MSF and MED. Consequently, this enables a more significant decrease in temperature per stage, ranging from 2 to 2.5 K, and an expansion in the possible number of MED or MSF steps from the typical 15 to 60. As a result, thermal desalination methods have significantly lower energy requirements, with a specific consumption rate decreasing from 32 kWhth/m^3 to 11 kWhth/m^3 (El-Ghonemy, 2012).

Energy recovery equipment, such as turbine or pressure exchanger systems, can achieve energy recovery in RO systems. Using a Pelton wheel or a turbocharger device, the turbine system transforms the brine's potential energy into mechanical energy (Chavan et al., 2007). This energy can be used to minimize the particular energy consumption in the system or as an auxiliary power source for the feed pump. Direct pressure transmission from brine to water supply is accomplished via pressure exchanger systems, also called work exchangers. With a maximum efficiency of 98%, pressure exchangers are the most efficient energy recovery devices currently available for use with brine. Compared to pressure exchangers, Pelton turbines and turbochargers have been around longer, but their energy recovery efficiencies are only 90%. Desalination-specific energy requirements were less than 1 kWh/m^3 for brackish waters and 2 to 4 kWh/m^3 for seawater when energy recovery devices were used (Nassrullah et al., 2020).

2.4 INTEGRATION WITH SUSTAINABLE ENERGY SOURCES

The significance of energy in our society is increasing to ensure a high standard of living and efficient operation of other components of our economy. Using energy resources is widely recognized as one of today's world's most significant and persistent challenges. Currently, around two billion people worldwide do not have access to electricity (Ahmed et al., 2020). The dependence on technology improved the quality of life in developed countries, and ongoing population growth in emerging nations undoubtedly led to increased energy demands. As more fossil fuels are burned for energy, the ozone layer thins, climate change happens, environmental problems worsen, and health risks for all living things on Earth rise. Nuclear resources, fossil fuels, and RES are the three primary categories among which energy sources can be classified. RES have the potential to play a large role in the future of the globe (Nwokediegwu et al., 2024). Energy can be replenished by using renewable sources, including wind, geothermal, solar, biomass, and a wide variety of other kinds of energy creation (M. Goosen et al., 2011). In order to generate energy that is devoid of greenhouse gases and other air pollutants, RES have the capacity to emit 0% or nearly 0% of these pollutant gasses. This allows them to provide energy that is free of these pollutants. Harvesting renewable energy in a decentralized manner to satisfy the energy requirements of rural and small-scale communities is a method that is dependable, economical, and environmentally sustainable. Even though many of these technologies are still in the process of being developed, there are a few renewable energy technologies that are currently being utilized in the twenty-first century (Ghaffour et al., 2013).

Sustainable water infrastructure increasingly relies on desalination technology to meet the rising demand for freshwater and address the developing problems with traditional water sources. Seawater or brackish water can be desalinized to remove salt and other contaminants, making it a significant resource for areas with water constraints. To reduce negative effects on the environment and ensure the process's continued success in the long run, desalination must be sustainable (Garg & Joshi, 2015; Manju & Sagar, 2017).

Desalination plants often use fossil fuels to generate electricity, releasing greenhouse gases and other pollutants exacerbating global warming and acid rain. As the Kyoto Protocol takes effect, numerous nations have committed to reducing greenhouse gas emissions (Mahi, 2001). Global per capita emissions are currently at 0.3 in developing nations, 5.5 in the United States of America, and 2.5 in Western Europe. The Kyoto Agreement mandates that these levels must be reduced to between 0.2 and 0.7 tons of carbon dioxide per capita per year. Furthermore, energy security and future sustainable development are becoming increasingly dependent on developing new and alternate energy sources due to the depletion of the limited fossil fuel supplies. The known petroleum reserves are projected to be exhausted at the current usage rate in under half a century.

2.5 RENEWABLE ENERGY INTEGRATED DESALINATION TECHNOLOGIES

In recent years, there has been a growing focus on the possible utilization of renewable energy as a clean and environmentally friendly power source for operating small-scale desalination plants in isolated communities. Incorporating RES, such as solar, wind, and geothermal power, into desalination systems holds great potential in the fight against water scarcity. This approach also offers a workable solution for mitigating the effects of climate change and water constraints (Ghaffour et al., 2015).

Wind, solar (either by photovoltaics or solar collectors), and geothermal are the three basic categories that can be used to classify desalination systems powered by RES

(García-Rodríguez, 2003). As shown in Figure 2.7, water production can be accomplished through thermal distillation or membrane desalination systems in conjunction with RES. There are instances in which these systems are connected to a conventional source of energy, such as a local power grid, to lessen the swings in energy and, consequently, the water production rate (Tzen & Morris, 2003).

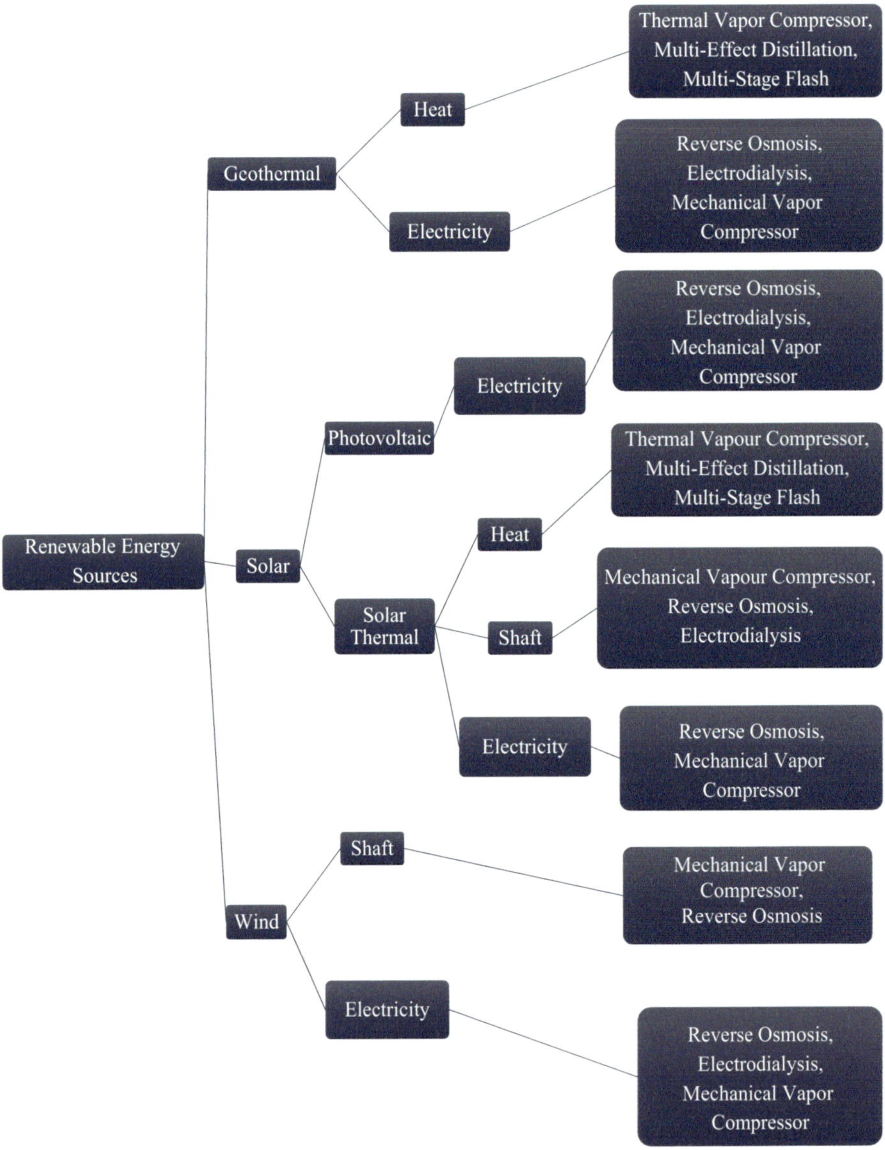

FIGURE 2.7 Possible integration of RES with desalination technology.

Figure 2.7 illustrates a possible combination of RES with desalination technology, such as solar, geothermal, and wind energy, utilized in two ways: as heat energy for thermal processes or as electrical energy for membrane-based processes.

2.5.1 Geothermal

Geothermal energy has a significant benefit compared to other RES in terms of its ability to offer a consistent and reliable energy supply (Gude, 2018). Wind and solar energy, in contrast, exhibit interruption, necessitating more complex capture mechanisms and expensive energy storage devices, which are also constrained in size, impeding the scalability of larger units. None of these adverse effects are present in the geothermal option. However, the drawbacks of targeting new geothermal reservoirs include the substantial expenses associated with discovery, the elevated risk of investment, and the significant expenditures involved in installation. However, the expensive nature of the desalination machine will be balanced out by the readily accessible geothermal heat created throughout its operation (Chandrasekharam, 2018). This is in contrast to heat generated from fossil fuels, which necessitates the continuous purchase of fuels unless waste heat can be utilized. Furthermore, there are other instances in which the significant expenses associated with exploration and the potential for financial loss might be mitigated (Prajapati et al., 2021).

In regions with access to suitable geothermal resources, geothermal heat provision can be advantageous because it is typically less expensive than solar power. However, economic modes must prove, case by case, which option is most beneficial (Almtairi et al., 2021). Hybrid systems that combine solar and geothermal power can also be considered. Using a liquid-liquid heat exchanger, geothermal heat can be directed into well-established, cost-effective technologies like MSF. Additionally, compared to MSF, the working temperature of geothermal heat with MED technology is lower, at less than 70°C. MSF and MED facilities are particularly competitive with less energy-intensive RO technology, which can be powered by inexpensive geothermal heat (Sohani et al., 2022).

2.5.2 Wind Energy

A wind turbine is used to convert the kinetic energy of air velocity into mechanical energy, which is then converted into electrical energy. This provides a means of harnessing wind energy. Wind power is an economically efficient and exceptionally eco-friendly kind of electricity generation (Ghaffour et al., 2013). The primary limitation of wind energy is its intermittent nature, which leads to fluctuations in energy generation due to the continually changing wind speeds. Furthermore, wind energy is predominantly accessible in distant regions, particularly along coastlines. To address this difficulty, it has been suggested and implemented to be integrated with a storage system, grid, or other RES (Lai et al., 2016). Wind power is commonly used to operate RO.

Tomaszewska et al. (2020) conducted techno-economic research on using wind energy in RO plants. They found that this method's cost of producing drinking water ranges from 2962 to 6457 $/m^3. Forstmeier et al. (2007) investigated the practicality of using MVC and wind-based RO. Meanwhile, Jin et al. (2019) researched the impact of wind intermittency on RO and proposed three potential solutions: energy storage, the integration of energy systems, and the temporary modification of RO capacity with a supplementary energy source. Rosales-Asensio et al. (2019) conducted an economic analysis to compare the cost-effectiveness of an RO process powered by wind energy to that of traditional energy sources. After using the levelized cost of freshwater as the critical factor, the researchers concluded that the wind-based RO technology has a lower levelized cost for producing freshwater.

2.5.3 Solar Energy

The process of utilizing semiconductors and solar collectors, both of which are components of photovoltaic cells, in order to extract energy from the sun falls under the category of solar energy. The sun emits radiant energy in the form of light and heat, which may be employed to generate both thermal and electrical energy. This energy can be released into the atmosphere. According to Delgado et al. (2020), solar energy may be divided into two basic categories: photovoltaic and solar thermal. According to this, solar energy can be classified. The solar thermal process involves using a solar collector to absorb solar energy. The solar energy that has been gathered is first converted into thermal energy, and then it is transmitted to a fluid that is moving through the absorber (Francisco et al., 2018). Through the process of turning heat energy into electrical and mechanical energy, it is possible to use heat energy either directly for thermal desalination or indirectly for membrane management systems. These methods are still not viable for the large-scale generation of considerable quantities of freshwater (Reif & Alhalabi, 2015). This is even though progress has been made in capturing solar energy and improving its application in water desalination systems. According to Wagner and Rubin (2014), solar thermal-powered desalination is restricted to areas that receive significant sunlight and require a comparatively big area for operation.

2.5.4 Photovoltaic Energy

A solar photovoltaic energy system operates by harnessing sunlight and turning it into electrical energy using a device that exhibits photovoltaic phenomena (Al-Karaghouli et al., 2010). The photovoltaic effect is the change in voltage when light radiation is applied to electrodes that have been built to be extremely effective, with a solid or liquid material positioned between them (Goetzberger et al., 2003). This transformation takes place when the electrodes are exposed to light radiation. Photovoltaic technology has emerged as a potentially advantageous option because it has a longer lifespan and lower prices for operations and capital expenditures. Nevertheless, obstacles such as the accumulation of dirt on solar panels decrease the efficiency of the process and restrict their energy output. Additionally, photovoltaic systems are typically utilized in areas with abundant solar radiation, which enhances their effectiveness but also presents a challenge due to the need for cooling. Weathering, dust, dependence on climatic variations, wind velocity, and humidity levels, particularly in hot regions, are prominent obstacles that have a noticeable impact on energy loss (Ghaffour et al., 2013; Ghaffour & Mujtaba, 2018).

2.6 DESALINATION TECHNOLOGIES AND BEST PRACTICES

Water supplies are often undervalued in many regions. However, water scarcity can pose significant obstacles to achieving sustainable development. The lack of available water affects the economy's advancement, weakens the well-being of individuals, contributes to the deterioration of the environment, and creates instability in politics and society (UNWWD, 2019). The implementation of strategies such as water conservation and the improvement of the efficiency of supply and distribution systems may provide some assistance; nonetheless, it is anticipated that their impact will be limited. Given that the global population is anticipated to increase by 50% by the year 2030, it is of the utmost importance to identify and cultivate new water resources to ensure the continuous and sustainable growth of future water supplies. International organizations such as the United Nations Development Program (UNDP), the World Health Organization (WHO) (WHO, 2011), and the United Nations Children's Fund (UNICEF), among others, have forecasted that 24 of the world's least developed countries, the majority of which are located in coastal areas with limited access to freshwater and electricity, will be required to make significant advancements in order to achieve the Millennium Development Goals (MDGs) in the areas of fundamental health, sanitation, and welfare (Gadhamshetty et al., 2014; Gude et al., 2010). For Middle Eastern countries that are both oil-rich and desert, desalination technology is essential for ensuring a steady supply of freshwater. However, several regions once abundant in water resources are experiencing significant water supply challenges. Water scarcity is becoming worse in countries with rapid population growth as a result of both climate change and shifting demography. This is particularly challenging due to the increased demand for freshwater caused by rapid industrialization. Seawater is a readily available solution to address the issue of water scarcity, particularly in coastal regions. Several case studies will be examined in this discussion, including those of the following: the water supply and desalination plants in the South-North water transportation project in the Kingdom of Saudi Arabia (Chandrasekharam et al., 2017), the sustainable water supply scheme in Singapore, the state of California in the United States of America (Bermudez-Contreras et al., 2008), and desalination plants in Australia (Heihsel et al., 2019).

Most water treated by desalination facilities in Florida consists of brackish groundwater or surface water rather than seawater. Several water plants in Florida get their source water from the coastal seawater that is located close to them. One of the state's only large-scale RO operation that uses a coastal surface water source is the Tampa Bay

Seawater Desalination facility. Currently, the Coquina Coast project, located in Flagler County, Northeast Florida, is evaluating to determine whether it is possible to adopt seawater desalination technology. Both land and innovative maritime techniques are currently being investigated. Texas is presently undergoing a severe drought, and it is anticipated that the accessibility of groundwater will diminish by 30% over the following five decades (California et al., 2008) (Table 2.2).

Additionally, it is anticipated that the population of Texas will increase by 21 million, which will result in an increased demand for renewable sources of freshwater. Due to the declining accessibility of surface and groundwater sources, Texas must investigate other methods for providing freshwater. Another option is to treat the abundant brackish groundwater utilizing desalination.

In 2005, the National Water Commission of the Australian Government reported that Australia is the driest inhabited continent in the world. This observation was made. The nation is facing catastrophic droughts, followed by forest fires and floods, continuously putting its water supply sources in peril. This is all due to climate change. The worst drought the continent has seen in a long time since the early 2000s was felt there. In 2006, Perth commissioned its first significant desalination plant, which generates 45 million m^3 annually. Due to the ongoing drought, a second desalination plant will be developed to double its capacity, increasing its annual production to around 50 million m^3 (Stover, 2007).

One facility was built in Queensland with a capacity of 48 million m^3, and another in Adelaide with a capacity of 100 million m^3 per annum. The most contentious and costly plant, with a capacity of 150 million m^3/d and an investment of $3.6 billion, was commissioned in Melbourne. However, now that the drought is less severe due to the recent heavy rains, not all plants are running continuously. As part of a larger strategy to lessen the severity of droughts and make better use of big reservoirs, the government has invested much money into water-saving, reusing, recycling, reclamation projects, and collecting rainfall. Australia is well-equipped to confidently deliver sustainable water supplies, thanks to its advanced infrastructure and management strategies, even though the current drought circumstances have passed (Williams et al., 2015).

According to Nexus (2018), Saudi Arabia's renewable water capacity is only 2400 $\times 10^6$ m^3, but the country still withdraws around 20,000 × 106 m^3 each year. With 25.7 million people, it is the biggest country on Earth without water bodies (Llamas and Custodio, 2003). The high salinity of the non-renewable groundwater sources accounts for almost 80% of the water supplies, necessitating desalination. Worldwide, 17% of desalination capacity is located in Saudi Arabia. It might cost as much as $200 billion for the nation to raise its daily desalination capacity by 6 × 10^3 m^3 (Mezher et al., 2011). The predicted yearly growth rate for desalination capacity was 9.5% from 2012 to 2015. In 2020, the Saline Water Conversion Corporation (SWCC) plans to invest $11.7 billion in capital expenditures and $4.5 billion in operating expenses. In addition to the development of desalination technology, this is a positive sign for companies that specialize in engineering, procurement, and construction.

In 2014, the nation of Saudi Arabia's Saline Water Conversion Corporation built the largest desalination facility in the world near the city of Ras Al-Khair, which is an industrial center. Water Technology 2015a reports that the plant has the capacity to generate

0.728 m^3 of water per day. Seventeen RO units and eight MSF units comprise the hybrid desalination plant structure. On a daily basis, the MSF units have the capacity to process 0.606 million m^3, while the RO units have the capacity to process 0.257 million m^3. The combined cycle power plant that is included in this project has a capacity of 2650 megawatts. It is estimated that the total cost of construction for the project was SAR 27 billion, which is about equivalent to $7.2 billion (Al-Mutaz & Wazeer, 2015). Between the years 2015 and 2016, it is anticipated that six large projects will be put up for application. The Jubail thermal and RO (hybrid) project is the most extensive undertaking of its kind (Baig & Al Kutbi, 1999). It is anticipated to be worth three billion dollars in the United States. Conversion plants for salty water that make use of RO technology are included in the five projects that are still in the plan. Rabigh Phase 4, which has a value of $1.2 billion; Umluj SWRO, which has a value of $30 million; Aqeer SWRO, which has a value of $20 million; and South Dhahran SWRO, which has a value of $200 million are the projects that are presented. Prioritizing advancements in engineering services, membrane and thermal desalination technologies, and the effective use of solar energy are all necessary steps in accomplishing this goal.

A low-temperature thermal desalination (LTTD) plant is on the Lakshadweep Islands. In May 2005, the National Institute of Ocean Technology (NIOT) established a land-based demonstration plant in Kavaratti under the Ministry of Earth Sciences (Article, 2014). This plant can produce 1 lakh liters of freshwater per day. The bathymetry of the ocean floor around the island was such that there was a water depth of 350 m at a distance of approximately 600 m from the shore. A temperature gradient of 15°C was applied, with the surface water temperature at 28°C and the water at a depth of 350m at 12°C. A 630 mm diameter and 600 m long high-density polyethylene (HDPE) pipes were used to extract cold water from a depth of around 350 m. At the land-based facility, the partitioned sump makes it easier for warm and cold water to move from the

TABLE 2.2 World's largest desalination plants (Reuters Staff, 2020)

SR. NO.	*NAME OF THE PLANT*	*METHOD USED*	*CAPACITY (M^3/DAY)*	*LOCATION*	*REFERENCE*
1	Ras Al-Khair	Thermal multi-stage flash & reverse osmosis	1,036,000	Saudi Arabia	Esmaeilion (2020)
2	Taweelah	Thermal multi-stage flash & reverse osmosis	230,000	UAF	JICA (2016)
3	Jubail Water and Power Company	Multi-effect distillation	23,500	Saudi Arabia	JICA (2016)
4	Umm Al Quwain	Reverse osmosis	590,000	UAE	JICA (2016)
7	Sorek-1	Reverse osmosis	624,000	Israel	Nagaraja Bhat (2016)
8	Sorek-2	Reverse osmosis	570,000	Israel	Nagaraja Bhat (2016)
9	Fujairah-2	Reverse osmosis	136,000	UAE	JICA (2016)

ocean to the infrastructure. The facility has operated consistently since then, meeting the demands of the local population of 10,000 people for more than four years. The salinity of the freshwater produced decreased from 35,000 parts per million (ppm) in the ocean to 280 ppm, below the acceptable level of 2000 ppm for drinking water (Joji, 2016).

Plant of desalination is located in Galdar at Bocabarranco beach, northwest (NW) of the island of Gran Canaria. Specifications of the Bocabarranco SWRO desalination facility: Production capacity: 7000 m^3 per day; Maximum recovery: 45% total dissolved solids; 100 milligrammes of H2O; Application: Domestic use. In addition to taking advantage of all the services offered by the infrastructure, the plant is a component of a bigger desalination system that covers an area of 10,000 m^2 (Shokri & Sanavi Fard, 2022). An adjacent wastewater treatment facility and an additional seawater RO desalination plant for agricultural use are located in separate buildings. The brine is sent to the shore via two pipes: one 300 mm diameter for the potable water plant's brine rejection and another 400 mm for agricultural purposes. Near the beach, both pipes empty. The decision was made to connect all emissions, including wastewater, through an extensive outfall that flows into the ocean (Nwokediegwu et al., 2024).

2.7 FUTURE DIRECTIONS AND RESEARCH OPPORTUNITIES

The sustainability of desalination systems can be evaluated based on four key components: environmental, economic, social, and technological. Technical feasibility and other major components are crucial in evaluating desalination procedures. These methods harm the environment because they use a lot of chemicals, money, and energy. Technology is crucial in reducing these effects. Factors that affect the economic impact include water availability and quality, energy source, process technology, labor costs, O&M expenditures, location, financing package, and the type of energy source used. Greenhouse gas emissions, which cause the Earth to warm and produce acid rain, are part of the environmental component. In addition, chemicals and brine are released into the environment, which can cause biological impacts in the areas that receive them, as well as eutrophication and a decline in water quality. The social variables encompass societal acceptability, trust in the amount and quality of water supplies, the need for water, acceptance of desalination technology, and trust in water providers (Goosen et al., 2016).

An important decline in the supply of drinking water brought about by population growth and industrialization has added to several major problems facing humanity. Therefore, the return on investment in creating new techniques for water desalination to address the global water crisis issues would be enormous. However, the operability and efficiency of various desalination systems mostly depend on an uninterrupted energy supply. The energy concept will, therefore, unavoidably play a crucial part in sustainable water desalination. Furthermore, it is crucial to prioritize using clean and sustainable energy sources to ensure the efficient operation of desalination systems and reduce the adverse environmental effects. This is because of the strict criteria enforced by policymakers and environmentalists, who seek to reduce the negative impacts of utilizing fossil fuels to operate desalination facilities. These impacts encompass the substantial release of greenhouse gases and several issues associated with desalination, such as membrane fouling, scaling, and brine disposal (Sadhwani et al., 2005). Incorporating RES into desalination systems can provide significant advantages as an ecologically sustainable option. Within the nexus of water and energy, the primary motivation behind endeavors is to ensure a sustainable future. An examination of the literature on water desalination revealed that RES can be effectively integrated with several sophisticated technologies to achieve satisfactory results. Although there have been substantial advancements in this area, the primary emphasis should be on maximizing specific techno-economic parameters to guarantee sustainable and efficient operations in the future (Shatat & Riffat, 2014).

Small-scale desalination facilities are operated by desalination methods powered by wind energy, which are not linked to a conventional power grid. These systems are highly efficient standalone microgrids specifically designed for this purpose. Significant efforts are still required to incorporate energy storage systems to optimize wind energy utilization and mitigate the intermittent nature of wind power (Komendantova, 2017).

Improving the efficiency of solar thermal energy applications in desalination systems, whether large or small, is crucial. This can be achieved through various means, such as increasing the capacity to store heat, developing better materials for solar collectors, and optimizing the surface area of devices that concentrate solar radiation through new technologies. Optimization of fluid solutions and materials for energy savings is still necessary in these domains. Concentrated solar power systems intended for drought-stricken regions must be sturdy enough to withstand intense heat and sandstorms. In addition, solar-dependent plants are more challenging to build and maintain since they require dynamic daylight tracking technology. This makes their manufacture and maintenance more expensive compared to flat-panel collectors (Ahmadvand et al., 2019).

Geothermal energy can provide substantial advantages for desalination systems in regions with sufficient geothermal resources, provided that an economic study is conducted for each specific situation to validate its viability. Moreover, developing desalination plants using geothermal energy is often feasible as long as the well has adequate capacity and no obstacles. Moreover, it is crucial to emphasize that geothermal heat sources with low enthalpy and cost-effectiveness can be extensively utilized as direct heat sources for many conventional thermal-dependent water desalination methods (Xevgenos et al., 2016).

2.8 CONCLUSION

This chapter thoroughly examines desalination technologies and their incorporation with RES. This statement emphasizes the crucial significance of desalination in addressing the worldwide water shortage, especially in dry and coastal areas. The chapter examines the progress made in thermal and membrane desalination technologies, the possible environmental consequences, and the significance of RES in mitigating the carbon emissions of desalination facilities. It emphasizes the need for innovative designs that utilize renewable energy to make desalination more environmentally friendly and economically viable. The chapter also presents case studies and future research opportunities, suggesting that advancements in engineering services, membrane technologies, and the harnessing of solar energy are essential for the sustainable development of desalination systems. The chapter concludes that desalination is a crucial technology for ensuring water security; its sustainability can be significantly enhanced by integrating it with RES, thereby contributing to a more sustainable future.

REFERENCES

Abdelkareem, M. A., El Haj Assad, M., Sayed, E. T., & Soudan, B. (2018). Recent progress in the use of renewable energy sources to power water desalination plants. *Desalination*, *435*, 97–113. https://doi.org/10.1016/j.desal.2017.11.018

Acwa Power Inaugurates Jubail 3a Iwp (Jazlah Water Desalination Company), the First Large-Scale Integrated Water Desalination - Solar Pv Project in Saudi Arabia. (2023). Acwa Power. https://www.acwapower.com/news/acwa-power-inaugurates-jubail-3a-iwp-jazlah-water-desalination-company/#:~:text=Jazlahencompasses a 45.5 MW,m3%2C using reverse osmosis technology

Aende, A., Gardy, J., & Hassanpour, A. (2020). Seawater desalination: A review of forward osmosis technique, its challenges, and future prospects. *Processes*, *8*(8). https://doi.org/10.3390/PR8080901

Ahmadvand, S., Abbasi, B., Azarfar, B., Elhashimi, M., Zhang, X., & Abbasi, B. (2019). Looking beyond energy efficiency: An applied review of water desalination technologies and an introduction to capillary-driven desalination. *Water (Switzerland)*, *11*(4). https://doi.org/10.3390/w11040696

Ahmed, F. E., Hashaikeh, R., & Hilal, N. (2020). Hybrid technologies: The future of energy efficient desalination – A review. *Desalination*, *495*(July), 114659. https://doi.org/10.1016/j.desal.2020.114659

Al-Amshawee, S., Yunus, M. Y. B. M., Azoddein, A. A. M., Hassell, D. G., Dakhil, I. H., & Hasan, H. A. (2020). Electrodialysis desalination for water and wastewater: A review. *Chemical Engineering Journal*, *380*(March 2019). https://doi.org/10.1016/j.cej.2019.122231

Al-Dousari, A., Al-Ghadban, A. N., & Sturchio, N. C. (2012). Marine environmental impacts of power-desalination plants in Kuwait. *Aquatic Ecosystem Health and Management*, *15*, 50–55. https://doi.org/10.1080/14634988.2012.667336

Al-Ghamdi, A. Y., Saraya, M. E. S. I., Al-Ghamdi, A. O., & Zabin, S. A. (2014). Study of physico-chemical properties of the surface and ground water. *American Journal of Environmental Sciences*, *10*(3), 219–235. https://doi.org/10.3844/AJESSP.2014.219.235

Alhaj, M., Hassan, A., Darwish, M., & Al-Ghamdi, S. G. (2017). A techno-economic review of solar-driven multi-effect distillation. *Desalination and Water Treatment*, *90*(January), 86–98. https://doi.org/10.5004/dwt.2017.21297

Al-Karaghouli, A., & Kazmerski, L. L. (2013). Energy consumption and water production cost of conventional and renewable-energy-powered desalination processes. *Renewable and Sustainable Energy Reviews*, *24*, 343–356. https://doi.org/10.1016/j.rser.2012.12.064

Al-Karaghouli, A., Renne, D., & Kazmerski, L. L. (2010). Technical and economic assessment of photovoltaic-driven desalination systems. *Renewable Energy*, *35*(2), 323–328. https://doi.org/10.1016/j.renene.2009.05.018

Alklaibi, A. M., & Lior, N. (2005). Membrane-distillation desalination: Status and potential. *Desalination*, *171*(2), 111–131. https://doi.org/10.1016/j.desal.2004.03.024

AlMadani, H. M. N. (2003). Water desalination by solar powered electrodialysis process. *Renewable Energy*, *28*(12), 1915–1924. https://doi.org/10.1016/S0960-1481(03)00014-4

Almtairi, A., Sharaf Eldean, M. A., Soliman, A. M., Mabrouk, A., & Fath, H. E. S. (2021). A new preliminary system design of using geothermal well brine heater for desalination/nanofiltration process. *Cleaner Engineering and Technology*, *4*(June), 100213. https://doi.org/10.1016/j.clet.2021.100213

Al-Mutaz, I. S., & Wazeer, I. (2015). Current status and future directions of MED-TVC desalination technology. *Desalination and Water Treatment*, *55*(1), 1–9. https://doi.org/10.1080/19443994.2014.910841

Ameri, M., Mohammadi, S. S., Hosseini, M., & Seifi, M. (2009). Effect of design parameters on multi-effect desalination system specifications. *Desalination*, *245*(1–3), 266–283. https://doi.org/10.1016/j.desal.2008.07.012

Andersen, I., & Bank, W. (2012). Environmental Impacts of Desalination. In *Renewable Energy Desalination: An Emerging Solution To Close the Water Gap in the Middle East and North Africa* (pp. 111–126).

Article, C. (2014). Case article desalination: Case study of Lakshadweep Islands and Gran Canaria Islands. *International Journal of Current Engineering and Technology*, *4*(2), 1073–1078.

Baig, M. B., & Al Kutbi, A. A. (1999). *Design features of a 20 migd SWRO desalination plant, Al Jubail, Saudi Arabia*. Water Supply. https://scholar.google.com/scholar?cluster=1958749401742212336&hl=en&as_sdt=0,5

Bamufleh, H., Abdelhady, F., Baaqeel, H. M., & El-Halwagi, M. M. (2017). Optimization of multi-effect distillation with brine treatment via membrane distillation and process heat integration. *Desalination*, *408*, 110–118. https://doi.org/10.1016/j.desal.2017.01.016

Belessiotis, V., Kalogirou, S., & Delyannis, E. (2016). Thermal Solar Desalination: Methods and Systems. In *Thermal Solar Desalination: Methods and Systems* (pp. 1–368). https://doi.org/10.1016/C2015-0-05735-5

Bermudez-Contreras, A., Thomson, M., & Infield, D. G. (2008). Renewable energy powered desalination in Baja California Sur, Mexico. *Desalination*, *220*(1–3), 431–440. https://doi.org/10.1016/j.desal.2007.01.046

Bhagwati, A., Shah, M., & Prajapati, M. (2023). Emerging technologies to sustainability: A comprehensive study on solar desalination for sustainable development. *Sustainable Manufacturing and Service Economics*, *2*, 100007. https://doi.org/10.1016/j.smse.2022.100007

Brower, M., Green, D., Hinrichs-Rahlwes, R., Sawyer, S., Sander, M., Taylor, R., Giner-Reichl, I., Teske, S., Lehmann, H., Alers, M., & Hales, D. (2014). *RENEWABLES 2014 GLOBAL STATUS REPORT Renewable Energy Policy Network for the 21st Century REN21*. https://scholar.google.com/scholar?hl=en&as_sdt=0%2C5&q=Adib%2C+R.%2C+Murdock%2C+H.%2C+Appavou%2C+F.%2C+Brown%2C+A.%2C+Epp%2C+B.%2C+Leidreiter%2C+A.%2C+Lins%2C+C.%2C+Murdock%2C+H.%2C+Musolino%2C+E.+and+Petrichenko%2C+K.%2C+2016.+Renewables+2016+Global+Sta

Bundschuh, J., Kaczmarczyk, M., Ghaffour, N., & Tomaszewska, B. (2021). State-of-the-art of renewable energy sources used in water desalination: Present and future prospects. *Desalination*, *508*(March). https://doi.org/10.1016/j.desal.2021.115035

California, I. B., Hernandez, S. E., & Ingenieria, T. De. (2008). *Using Coastline Geothermal Resources for Desalination. October*, 1–5.

Chandrasekharam, D. (2018). Water for the millions: focus Saudi Arabia. *Water-Energy Nexus*. https://www.sciencedirect.com/science/article/pii/S2588912518300304

Chandrasekharam, D., Lashin, A., Al Arifi, N., Al Bassam, A., & Varun, C. (2017). Desalination of Seawater using Geothermal Energy to Meet Future Fresh Water Demand of Saudi Arabia. *Water Resources Management*, *31*(3), 781–792. https://doi.org/10.1007/s11269-016-1419-2

Chavan, P. V., Dennett, K. E., Marchand, E. A., & Gustin, M. S. (2007). Evaluation of small-scale constructed wetland for water quality and Hg transformation. *Journal of Hazardous Materials*, *149*(3), 543–547. https://doi.org/10.1016/j.jhazmat.2007.06.077

Chekifi, T., Benmoussa, A., & Boukraa, M. (2024). Desalination Powered by Renewables: A Challenge and an AI Opportunity. In *Water Resources Management* (*Issue 0123456789*). Netherlands: Springer. https://doi.org/10.1007/s11269-024-03935-4

Cui, C., Xi, Z., Liu, S., & Sun, J. (2019). An enumeration-based synthesis framework for multi-effect distillation processes. *Chemical Engineering Research and Design*, *144*, 216–227. https://doi.org/10.1016/j.cherd.2019.02.018

Darre, N. C., & Toor, G. S. (2018). Desalination of Water: A Review. *Current Pollution Reports*, *4*(2), 104–111. https://doi.org/10.1007/s40726-018-0085-9

Darwish, M. A. (1991). Thermal analysis of multi-stage flash desalting systems. *Desalination*, *85*(1), 59–79. https://doi.org/10.1016/0011-9164(91)85147-M

Delgado, W. R., Beach, T., & Luzzadder-Beach, S. (2020). Solar desalination: Cases, synthesis, and challenges. *Wiley Interdisciplinary Reviews: Water*, *7*(3), e1434. https://doi.org/10.1002/wat2.1434

Edzwald, J. K., & Haarhoff, J. (2011). Seawater pretreatment for reverse osmosis: Chemistry, contaminants, and coagulation. *Water Research*, *45*(17), 5428–5440. https://doi.org/10.1016/j.watres.2011.08.014

El-Ghonemy, A. M. K. (2012). Future sustainable water desalination technologies for the Saudi Arabia: A review. *Renewable and Sustainable Energy Reviews*, *16*(9), 6566–6597. https://doi.org/10.1016/j.rser.2012.07.026

Elimelech, M., & Phillip, W. A. (2011). The future of seawater desalination: Energy, technology, and the environment. *Science*, *333*(6043), 712–717. https://doi.org/10.1126/science.1200488

Engelien, H. K., & Skogestad, S. (2005). Multi-effect distillation applied to an industrial case study. *Chemical Engineering and Processing: Process Intensification*, *44*(8), 819–826. https://doi.org/10.1016/j.cep.2004.06.015

Esmaeilion, F. (2020). Hybrid renewable energy systems for desalination. *Applied Water Science*, *10*(3), 1–47. https://doi.org/10.1007/s13201-020-1168-5

Feria-Díaz, J. J., Correa-Mahecha, F., López-Méndez, M. C., Rodríguez-Miranda, J. P., & Barrera-Rojas, J. (2021). Recent desalination technologies by hybridization and integration with reverse osmosis: A review. *Water (Switzerland)*, *13*(10), 1–40. https://doi.org/10.3390/w13101369

Forstmeier, M., Mannerheim, F., D'Amato, F., Shah, M., Liu, Y., Baldea, M., & Stella, A. (2007). Feasibility study on wind-powered desalination. *Desalination*, *203*(1–3), 463–470. https://doi.org/10.1016/J.DESAL.2006.05.009

Francisco, F., Leijon, J., Boström, C., Engström, J., & Sundberg, J. (2018). Wave power as solution for off-grid water desalination systems: Resource characterization for Kilifi-Kenya. *Energies*, *11*(4), 1004. https://doi.org/10.3390/en11041004

G., M., F., S., J., N., & A., P. (2011). Optimization of Hybrid Desalination Processes Including Multi Stage Flash and Reverse Osmosis Systems. *Desalination, Trends and Technologies*. https://doi.org/10.5772/15048

Gadhamshetty, V., Gude, V. G., & Nirmalakhandan, N. (2014). Thermal energy storage system for energy conservation and water desalination in power plants. *Energy, 66*, 938–949. https://doi.org/10.1016/j.energy.2014.01.046

García-Rodríguez, L. (2003). Renewable energy applications in desalination: State of the art. *Solar Energy, 75*(5), 381–393. https://doi.org/10.1016/j.solener.2003.08.005

Garcia-Rodriguez, L., Palmero-Marreroa, A. I., & Gbmez-Camachob, C. (2002). Comparison of solar thermal technologies for applications in seawater desalination. *Elsevier, 142*, 135–142. https://www.sciencedirect.com/science/article/pii/S0011916401004325

Garg, M. C., & Joshi, H. (2015). Optimization and economic analysis for a small scale nanofiltration and reverse osmosis water desalination system. *Water Science and Technology: Water Supply, 15*(5), 1027–1033. https://doi.org/10.2166/ws.2015.059

Ghaffour, N., Bundschuh, J., Mahmoudi, H., & Goosen, M. F. A. (2015). Renewable energy-driven desalination technologies: A comprehensive review on challenges and potential applications of integrated systems. *Desalination, 356*, 94–114. https://doi.org/10.1016/j.desal.2014.10.024

Ghaffour, N., Lattemann, S., Missimer, T., Choon Ng, K., & Amy, G. (2013). Seawater desalination using renewable energy: solar, geothermal, and wind 1 2. In *Proceeding of International Conference on Applied Energy (ICAE 2013).*

Ghaffour, N., & Mujtaba, I. M. (2018). Desalination using renewable energy. In *Desalination* (Vol. 435, Issue 96). Center for Renewable Energy Sources. https://doi.org/10.1016/j.desal.2018.01.029

Goetzberger, A., Hebling, C., & Schock, H. (2003). Photovoltaic materials, history, status and outlook. *Materials Science and Engineering: R: Reports.* https://www.sciencedirect.com/science/article/pii/S0927796X0200092X

Goh, P. S., Matsuura, T., Ismail, A. F., & Ng, B. C. (2017). The Water–Energy Nexus: Solutions towards Energy-Efficient Desalination. *Energy Technology, 5*(8), 1136–1155. https://doi.org/10.1002/ENTE.201600703

Gooch-Moore, J., Goodwin, K. D., Dorsey, C., Ellender, R. D., Mott, J. B., Ornelas, M., Sinigalliano, C., Vincent, B., Whiting, D., & Wolfe, S. H. (2011). New USEPA water quality criteria by 2012: GOMA concerns and recommendations. *Iwaponline.Com, 9*(4). https://doi.org/10.2166/wh.2011.156

Goosen, M., Mahmoudi, H., Ghaffour, N. et al. (2011). Application of Renewable Energies for Water Desalination. *Desalination, Trends and Technologies.* https://doi.org/10.5772/14011

Goosen, M. F. A., Mahmoudi, H., Ghaffour, N., Bundschuh, J., & Yousef, Y. A. (2016). *A critical evaluation of renewable energy technologies for desalination.* 233–258. https://doi.org/10.1142/9789813141124_0032

Gude, V. G. (2016). Desalination and sustainability - An appraisal and current perspective. *Water Research, 89*, 87–106. https://doi.org/10.1016/j.watres.2015.11.012

Gude, V. G. (2018). Geothermal Source for Water Desalination-Challenges and Opportunities. In *Renewable Energy Powered Desalination Handbook: Application and Thermodynamics* (Vol. *1*). Elsevier Inc. https://doi.org/10.1016/B978-0-12-815244-7.00004-0

Gude, V. G., Nirmalakhandan, N., & Deng, S. (2010). Renewable and sustainable approaches for desalination. *Renewable and Sustainable Energy Reviews, 14*(9). https://doi.org/10.1016/j.rser.2010.06.008

Hatzikioseyian, A., Vidali, R., & Kousi, P. (2003). Modeling and thermodynamic analysis of a Multi-Effect Distillation (MED) plant for seawater desalination. *National Technical University of Athens Journal (NTUA), January 2003.*

Heihsel, M., Lenzen, M., Malik, A., & Geschke, A. (2019). The carbon footprint of desalination: An input-output analysis of seawater reverse osmosis desalination in Australia for 2005–2015. *Desalination, 454*, 71–81. https://doi.org/10.1016/j.desal.2018.12.008

Huang, Z. H., Yang, Z., Kang, F., & Inagaki, M. (2017). Carbon electrodes for capacitive deionization. *Journal of Materials Chemistry A*, *5*(2), 470–496. https://doi.org/10.1039/c6ta06733f

Hussain, A., Arif, S. M., & Aslam, M. (2017). Emerging renewable and sustainable energy technologies: State of the art. *Renewable and Sustainable Energy Reviews*, *71*(June 2015), 12–28. https://doi.org/10.1016/j.rser.2016.12.033

Ihsanullah, I., Atieh, M. A., Sajid, M., & Nazal, M. K. (2021). Desalination and environment: A critical analysis of impacts, mitigation strategies, and greener desalination technologies. *Science of the Total Environment*, *780*, 146585. https://doi.org/10.1016/j.scitotenv.2021.146585

Intelligence, G. (2011). Global water intelligence. *Globalwaterintel.Com*. https://www.globalwaterintel.com/client_media/uploaded/GWI%20Media%20pack%20A4%2014_15(Screenversion).pdf

JICA. (2016). *The Survey on Feasibility of Desalination Projects in Sub-Saharan Africa Final Report. June*. https://openjicareport.jica.go.jp/pdf/12264560.pdf

Jin, G., Mo, Y., Li, M., Tang, H. et al. (2019). Desalinization and salinization: A review of major challenges for coastal reservoirs. *Journal of Coastal Research*. https://meridian.allenpress.com/jcr/article-abstract/35/3/664/9995

Joji, V. S. (2016). Sustainable water development and suitability of ground water for irrigation purposes in a small coral Island of Minicoy, Union Territory of Lakshadweep, India. *Journal of the Indian Society of Coastal Agricultural Research*, *34*(1), 97–109. https://epubs.icar.org.in/index.php/JISCAR/article/download/92722/37464/237995

Jones, E., Qadir, M., van Vliet, M. T. H., Smakhtin, V., & Kang, S. mu. (2019). The state of desalination and brine production: A global outlook. *Science of the Total Environment*, *657*, 1343–1356. https://doi.org/10.1016/j.scitotenv.2018.12.076

Kalogirou, S. A. (2005). Seawater desalination using renewable energy sources. *Progress in Energy and Combustion Science*, *31*(3), 242–281. https://doi.org/10.1016/j.pecs.2005.03.001

Komendantova, N. (2017). Renewable energy policy and mitigating the risks for investment. In *Renewable Energy Technologies for Water Desalination* (pp. 235–250). https://doi.org/10.1201/9781315643915

Kress, N. (2019). *Marine Impacts of Seawater Desalination: Science, Management, and Policy*. https://doi.org/10.1016/C2016-0-01997-6

Kucera, J. (2014). Desalination: Water from Water. *Desalination: Water from Water*, *9781118208*, 1–624. https://doi.org/10.1002/9781118904855

Lai, W., Ma, Q., Lu, H., Weng, S., Fan, J., & Fang, H. (2016). Effects of wind intermittence and fluctuation on reverse osmosis desalination process and solution strategies. In *Desalination* (Vol. *395*, pp. 17–27). https://doi.org/10.1016/j.desal.2016.05.019

Lawal, D., Antar, M., Khalifa, A., Zubair, S., & Al-Sulaiman, F. (2018). Humidification-dehumidification desalination system operated by a heat pump. *Energy Conversion and Management*, *161*, 128–140. https://doi.org/10.1016/j.enconman.2018.01.067

Lee, K. P., Arnot, T. C., & Mattia, D. (2011). A review of reverse osmosis membrane materials for desalination-Development to date and future potential. In *Journal of Membrane Science* (Vol. *370*, Issues 1–2, pp. 1–22). Elsevier. https://doi.org/10.1016/j.memsci.2010.12.036

Lian-Ying, W., Sheng-Nan, X., & Cong-Jie, G. (2012). Simulation of Multi-stage Flash(MSF) Desalination Process. *Advances in Materials Physics and Chemistry*, *2*(4), 200–205. https://doi.org/10.4236/ampc.2012.24b052

Llamas, M., & Custodio, E. (2003). Intensive use of groundwater: a new situation which demands proactive action. *Fundacionbotin.Org*. https://fundacionbotin.org/89dguuytdfr276ed_uploads/Observatorio%20Tendencias/PUBLICACIONES/LIBROS%20SEM%20INTERN/intensive%20use%20groundwater/cap01-intensiveuse.pdf

Loganathan, K., Sulaiti, H. Al et al. (2019). Distribution of salinity and trace elements in surface seawater of the Arabian Gulf surrounding the State of Qatar. *Desalination and Water Treatment*. https://www.sciencedirect.com/science/article/pii/S1944398624154768

Ludwig, H. (2022). *Reverse Osmosis Seawater Desalination Volume 1* (Vol. *1*). https://doi.org/10.1007/978-3-030-81931-6

Mabrouk, A. N., & Fath, H. E. S. (2015). Technoeconomic study of a novel integrated thermal MSF-MED desalination technology. *Desalination*, *371*, 115–125. https://doi.org/10.1016/j.desal.2015.05.025

Mahi, P. (2001). Developing environmentally acceptable desalination projects. *Desalination*. https://www.sciencedirect.com/science/article/pii/S0011916401002600

Manju, S., & Sagar, N. (2017). Renewable energy integrated desalination: A sustainable solution to overcome future fresh-water scarcity in India. *Renewable and Sustainable Energy Reviews*, *73*, 594–609. https://doi.org/10.1016/j.rser.2017.01.164

Mavukkandy, M. O., Chabib, C. M., Mustafa, I., Al Ghaferi, A., & AlMarzooqi, F. (2019). Brine management in desalination industry: From waste to resources generation. *Desalination*, *472*. https://doi.org/10.1016/j.desal.2019.114187

Mbarga, A. A., Song, L., Ross Williams, W., & Rainwater, K. (2014). Integration of renewable energy technologies with desalination. *Current Sustainable/Renewable Energy Reports*, *1*(1), 11–18. https://doi.org/10.1007/s40518-013-0002-1

Mezher, T., Fath, H., Abbas, Z., & Khaled, A. (2011). Techno-economic assessment and environmental impacts of desalination technologies. *Desalination*, *266*(1–3), 263–273. https://doi.org/10.1016/j.desal.2010.08.035

Miller, S., Shemer, H., & Semiat, R. (2015). Energy and environmental issues in desalination. *Desalination*, *366*, 2–8. https://doi.org/10.1016/j.desal.2014.11.034

Mohammadzadeh Bina, S., Jalilinasrabady, S., & Fujii, H. (2018). Exergoeconomic analysis and optimization of single and double flash cycles for Sabalan geothermal power plant. *Geothermics*, *72*, 74–82. https://doi.org/10.1016/j.geothermics.2017.10.013

Nafey, A. S. M. T. (1988). *Design and simulation of seawater thermal desalination plants. November.*

Nafey, A. S., Fath, H. E. S., & Mabrouk, A. A. (2006). Thermo-economic investigation of multi effect evaporation (MEE) and hybrid multi effect evaporation-multi stage flash (MEE-MSF) systems. *Desalination*, *201*(1–3), 241–254. https://doi.org/10.1016/j.desal.2005.09.044

Nagaraja Bhat. (2016). *Reverse Osmosis Monitoring & Optimization (ROMO): SAP HANA IoT Solution for Water Treatment | SAP Blogs*. https://blogs.sap.com/2016/03/15/romo-reverse-osmosis-monitoring-optimization-a-sap-hana-solution-for-reverse-osmosis-water-treatment/

Nassrullah, H., Anis, S. F., Hashaikeh, R., & Hilal, N. (2020). Energy for desalination: A state-of-the-art review. *Desalination*, *491*(June). https://doi.org/10.1016/j.desal.2020.114569

Nelson, V. C. (2011). Introduction To Renewable Energy. *Introduction to Renewable Energy*, 1–351. https://doi.org/10.1201/9781439891209

Nexus, D. C.-W.-E., & 2018, undefined. (2018). *Water for the millions: focus Saudi Arabia. Elsevier*. https://www.sciencedirect.com/science/article/pii/S2588912518300304

Nwokediegwu, Z. Q. S., Obaigbena, A., Majemite, M. T., Daraojimba, O. H., Oliha, J. S., & Dada, M. A. (2024). Review of innovative approaches in water infrastructure: Sustainable desalination and public-private partnerships. *International Journal of Science and Research Archive*, *11*(1), 324–332. https://doi.org/10.30574/ijsra.2024.11.1.0072

Panagopoulos, A. (2021). Water-energy nexus: desalination technologies and renewable energy sources. *Environmental Science and Pollution Research*, *28*(17), 21009–21022. https://doi.org/10.1007/s11356-021-13332-8

Petersen, K. L., Frank, H., Paytan, A., & Bar-Zeev, E. (2018). Impacts of Seawater Desalination on Coastal Environments. *Sustainable Desalination Handbook: Plant Selection, Design and Implementation*, 437–463. https://doi.org/10.1016/B978-0-12-809240-8.00011-3

Pouyfaucon, A. B., & García-Rodríguez, L. (2018). Solar thermal-powered desalination: A viable solution for a potential market. In *Desalination* (Vol. *435*, pp. 60–69). https://doi.org/10.1016/j.desal.2017.12.025

Prajapati, M., Shah, M., & Soni, B. (2022). A comprehensive review of the geothermal integrated multi-effect distillation (MED) desalination and its advancements. *Groundwater for Sustainable Development, 19*(July), 100808. https://doi.org/10.1016/j.gsd.2022.100808

Prajapati, M., Shah, M., Soni, B., Parikh, S., Sircar, A., Balchandani, S., Thakore, S., & Tala, M. (2021). Geothermal-solar integrated groundwater desalination system: Current status and future perspective. *Groundwater for Sustainable Development, 12*(October 2020), 100506. https://doi.org/10.1016/j.gsd.2020.100506

Reif, J. H., & Alhalabi, W. (2015). Solar-thermal powered desalination: Its significant challenges and potential. *Renewable and Sustainable Energy Reviews, 48*, 152–165. https://doi.org/10.1016/j.rser.2015.03.065

Reuters Staff. (2020). *Inside the world's biggest water desalination plants | Reuters*. https://www.reuters.com/article/us-saudi-water-desalination-idUSKBN26Y1HD

Rosales-Asensio, E., Borge-Diez, D., Pérez-Hoyos, A., & Colmenar-Santos, A. (2019). Reduction of water cost for an existing wind-energy-based desalination scheme: A preliminary configuration. *Energy*. https://www.sciencedirect.com/science/article/pii/S0360544218322084

Sadhwani, J. J., Veza, J. M., & Santana, C. (2005). Case studies on environmental impact of seawater desalination. *Desalination, 185*(1–3), 1–8. https://doi.org/10.1016/j.desal.2005.02.072

Shahzad, M. W., Burhan, M., Ang, L., & Ng, K. C. (2017). Energy-water-environment nexus underpinning future desalination sustainability. In *Desalination* (Vol. *413*, pp. 52–64). https://doi.org/10.1016/j.desal.2017.03.009

Shatat, M., & Riffat, S. B. (2014). Water desalination technologies utilizing conventional and renewable energy sources. *International Journal of Low-Carbon Technologies, 9*(1), 1–19. https://doi.org/10.1093/ijlct/cts025

Shokri, A., & Sanavi Fard, M. (2022). A sustainable approach in water desalination with the integration of renewable energy sources: Environmental engineering challenges and perspectives. *Environmental Advances, 9*(April), 100281. https://doi.org/10.1016/j.envadv.2022.100281

Sohani, A., Delfani, F., Hosseini, M., Sayyaadi, H., Karimi, N., Li, L. K. B., & Doranehgard, M. H. (2022). Dynamic multi-objective optimization applied to a solar-geothermal multi-generation system for hydrogen production, desalination, and energy storage. *International Journal of Hydrogen Energy, 47*(74), 31730–31741. https://doi.org/10.1016/j.ijhydene.2022.03.253

Stover, R. L. (2007). *Energy efficiency in Perth's desalination plant.* https://scholar.google.com/scholar?hl=en&as_sdt=0%2C5&q=Energy+efficiency+in+Perth%27s+desalination+plant&btnG=

Subramani, A., & Jacangelo, J. G. (2015). Emerging desalination technologies for water treatment: A critical review. *Water Research, 75*, 164–187. https://doi.org/10.1016/j.watres.2015.02.032

Tomaszewska, B., Bundschuh, J., Pająk, L., Dendys, M., Delgado Quezada, V., Bodzek, M., Armienta, M. A., Muñoz, M. O., & Kasztelewicz, A. (2020). Use of low-enthalpy and waste geothermal energy sources to solve arsenic problems in freshwater production in selected regions of Latin America using a process membrane distillation – Research into model solutions. *Science of the Total Environment, 714*, 136853. https://doi.org/10.1016/j.scitotenv.2020.136853

Tzen, E., & Morris, R. (2003). Renewable energy sources for desalination. *Solar Energy, 75*(5), 375–379. https://doi.org/10.1016/j.solener.2003.07.010

UNWWD. (2019). Leaving no one behind: the United Nations World Water Development. *WWAP (UNESCO World Water Assessment Programme)*, 186. https://doi.org/.1037//0033-2909.I26.1.78

Vishwanath Kumar, P., Kumar, A., Prakash, O., & Kaviti, A. K. (2015). Solar stills system design: A review. *Renewable and Sustainable Energy Reviews, 51*, 153–181. https://doi.org/10.1016/j.rser.2015.04.103

Voutchkov, N., & Kaiser, G. (2019). Of Desalination Plant Concentrate-Desalination Industry Position Paper–Energy and Environment Committee of the International Desalination Association .... *Researchgate.Net*. https://www.researchgate.net/publication/336879311

Wagner, S., & Rubin, E. S. (2014). Economic implications of thermal energy storage for concentrated solar thermal power. *Renewable Energy*. https://doi.org/10.1016/j.renene.2012.08.013

WHO. (2011). Safe Drinking-water from Desalination. *World Health Organization, WHO/HSE/WS*(1), 34. http://apps.who.int/iris/bitstream/10665/70621/1/WHO_HSE_WSH_11.03_eng.pdf

Williams, P. M. et al. (2015). Technology for freeze concentration in the desalination industry. *Desalination*. https://www.sciencedirect.com/science/article/pii/S0011916414005463

van Wyk, S., van der Ham, A., & Kersten, S. R. A. (2020). Analysis of the energy consumption of supercritical water desalination (SCWD). *Desalination*. https://www.sciencedirect.com/science/article/pii/S0011916419314304

Xevgenos, D., Moustakas, K., Malamis, D., & Loizidou, M. (2016). An overview on desalination & sustainability: Renewable energy-driven desalination and brine management. *Desalination and Water Treatment, 57*(5), 2304–2314. https://doi.org/10.1080/19443994.2014.984927

Geothermal Energy Integrated Desalination

3

3.1 INTRODUCTION

The primary factors contributing to water scarcity are pollution, climate change, rising human needs, and excessive water usage (Damania et al., 2017). The issue of freshwater scarcity is particularly severe in certain areas of the world, while an increasing number of countries are facing drought conditions due to global warming. The significance of water and its natural resources in sustaining life on Earth is widely accepted (Panagopoulos, 2021). Water is crucial for sustaining all life forms and is essential in various domains such as human survival, agriculture, and industrial processes (Fewtrell, 2013). Nevertheless, numerous countries, particularly developing ones, have experienced significant water scarcity issues throughout the past few decades. A vital component of agricultural output, water also has a crucial function in ensuring food security. On a global scale, the agriculture and food sectors dominate 80–90% of the world's freshwater resources, with 70% exclusively allocated for irrigation purposes (Koroneos & Roumbas, 2012).

Some remote locations worldwide, such as arid coastal parts of the Middle East and specific islands in the Mediterranean and Caribbean, are now facing a critical shortage of drinkable water. According to recent information from United Nations (UN) sources, over one billion people worldwide face a critical shortage of clean water, which is increasingly becoming a matter of life and death (Darre & Toor, 2018).

According to the study, the regions experiencing the most severe water shortages are those located in Western Asia and Northern Africa (Arab countries) (Dawoud, 2012). Despite the scarcity of water, Arab countries have had substantial population growth, accompanied by a noticeable rise in the agricultural industry throughout the past three decades. According to UN projections, the population of the 22 Arab countries is expected to reach 459.23 million by 2020, compared to the 281.22 million population estimated in 2000. By 2025, we anticipate that around 60% of the world's population will be experiencing severe water shortages. In addition, 30% of all fatalities and

DOI: 10.1201/9781003530893-3

80–90% of all illnesses in underdeveloped nations are attributed to the widespread consumption of water that is harmful to health (El-Ghonemy, 2012).

During the Industrial Revolution and until now, fossil fuels, specifically oil, coal, and natural gas, have been the primary energy sources (Gude, 2018). Utilizing these fossil fuels has significantly contributed to the acceleration of technological and societal advancement. Currently, we are confronted with local and global obstacles in ecology, energy, and economics. We must transition away from fossil fuels toward other sources that are more environmentally friendly and sustainable. Renewable energy sources, such as geothermal, biomass, solar, hydro, wind, and ocean, are viewed as highly viable options for completely replacing our current energy sources (Prajapati et al., 2021b). Several entities, such as corporations, governments, organizations, and individuals, are responsible for achieving a successful and efficient energy transition encompassing the entire global energy infrastructure. Renewable energy is becoming increasingly important for a variety of reasons (Ghaffour et al., 2013). There are many reasons to switch to renewable energy, but climate change is one of the most pressing and needs immediate attention. According to recent indications, unless drastic measures are taken by the middle of the twenty-first century, the planet will likely become inhabitable. We must complete the energy transition by phasing out fossil fuels and replacing them with renewables and other low-carbon options to reach our net-zero or carbon-neutral energy targets. It would be wise to examine the present energy sources to see what needs doing before moving on with specific renewable energy solutions (Almtairi et al., 2021).

Desalination is a widely employed process for producing freshwater by eliminating minerals, pollutants, and salts from salt water (Shelare et al., 2023). The primary barrier to desalination has been the expenditure of energy. Desalination necessitates far more energy than typical water treatment procedures. Consequently, the desalination process incurs greater expenses and generates increased levels of greenhouse gas emissions. Seawater desalination's energy usage accounts for 50%–70% of the total operational cost. Conventional desalination plants rely on evaporating seawater and then condensing the resulting humid air using thermal or electrical power from traditional sources. The production of 1000 m^3 per day of freshwater for a total of 10 kilotons per year necessitates a substantial energy input, usually in the range of millions of tons of oil equivalent or kilotons of oil equivalent (ktoe) (Prajapati, Shah, & Soni, 2021a). The current rise in popularity of novel systems driven by renewable sources can be attributed to this factor. Over time, desalination systems have developed and become more efficient, and renewable energy sources have started utilizing them as an effective option to address the lack of water and electricity. The primary sources of desalination are solar, Geothermal, and wind energy (Aviña-Jiménez et al., 2016).

The key objective of the research presented in this chapter is to explore the viability and effectiveness of integrating geothermal energy with desalination processes to produce freshwater. The research aims to address the critical issue of freshwater scarcity, particularly in regions suffering from severe potable water shortages, by leveraging the consistent and renewable nature of geothermal energy. By examining various geothermal desalination systems, including membrane, thermal, and hybrid systems, the study seeks to demonstrate how geothermal energy can be a sustainable and environmentally friendly solution for desalination, potentially revolutionizing water resource management and supporting sustainable development in water-stressed areas. The research delves into practical applications, case studies, and the global potential of geothermal desalination, highlighting its capacity to generate freshwater without the significant energy expenditure and greenhouse gas emissions associated with conventional desalination methods. Section 3.2 addresses desalination among other uses for geothermal energy as a viable and ecologically benign substitute for fossil fuels. It emphasizes how geothermal resources are categorized depending on temperature and enthalpy levels; hydrothermal fields are the most often occurring type. The section also outlines the applications of geothermal energy, such as direct heating and electricity generation, through different power cycles like the Organic Rankine Cycle (ORC). Section 3.3 discusses the innovative concept of using geothermal water for desalination. It emphasizes the low-temperature sources for desalination operations and the use of high-temperature geothermal sources for power generation. The section provides an overview of different desalination systems, such as membrane, thermal, and hybrid, that can be powered by geothermal energy. The section also outlines the advantages of geothermal desalination, including its potential for simultaneous electricity and water production, high-capacity factor, suitability for low-temperature desalination, and environmental benefits. Subsequent sections describe case studies and worldwide potential for geothermal desalination as subjects for additional investigation. Section 3.4 discusses case studies on geothermal desalination in various locations, including Mexico, Greece, India, and the United States. It highlights the unique geothermal resources of each region and their potential for desalination. The section details the implementation of different desalination methods, such as multi-effect distillation (MED) and reverse osmosis (RO), powered by geothermal energy. The case studies provide insights into the practical applications and challenges of integrating geothermal energy into desalination processes.

3.2 GEOTHERMAL ENERGY

Geothermal energy is the thermal energy in hot fluid, called geofluid, that is found in the Earth's crust and can exist as vapor, liquid, or a mixture of both (Systems, 2006). Several factors, such as the depth of the geothermal well, the temperature of the geofluid, the amount of accessible geofluid, and the chemical composition of the rock formations surrounding the well, impact the varying quality of the geothermal resource across different sites (Bora, 2016). The classification of geothermal resources is determined by their temperature, more particularly their enthalpy value. Hydrothermal fields are the most plentiful geothermal resources, characterized by medium enthalpy and being primarily composed of water (as geofluid) (Sircar et al., 2017; Stefansson, 2000). Conversely, high-enthalpy fields are predominantly composed of steam. High-enthalpy sources are characterized by temperatures that surpass 200°C and are commonly located in volcanic regions and along island chains. Enthalpy sources of moderate magnitude have temperatures within the range of between 150°C and 200°C, while sources of low magnitude have temperatures under 150°C (Green et al., 2021; Puppala & Jha, 2021).

Both hydrothermal and enhanced geothermal systems are considered the two distinct types of geothermal sources. Naturally occurring hydrothermal sources, which contain thermofluid that can be pumped around and exploited above ground, are available. However, enhanced geothermal systems are a more recent technology employed in situations with abundant hot rock but insufficient water for utilization. In these conditions, water is pumped, and a reservoir is created in enhanced geothermal systems (DiPippo, 2016; Goosen et al., 2010).

Geothermal energy has a wide range of applications, as it harnesses heat that can be utilized in various processes to produce valuable outputs (Rybach & Kohl, 2018). Unlike power generation systems, space heating is just one example that doesn't necessitate a significant investment. The temperature of the geothermal source plays a crucial role in determining the most appropriate useful outputs and processes (Chandrasekharam & Bundschuh, 2008). Space heating can be achieved using geothermal sources, including just a few meters underground. During winter, underground soil temperatures tend to be warmer than the surrounding air. Heat pumps can efficiently harness heat from the ground, resulting in a higher coefficient of performance and lower power consumption than air-source heat pumps. It is worth mentioning that the use of geothermal heat pumps has been steadily rising over the years. Additional important applications involve swimming, desalination, agriculture, and bathing (Sircar et al., 2022).

The utilization of geothermal energy can be exploited either directly for the purpose of heating, such as in the case of district heating, or indirectly for the generation of electricity. When it comes to geothermal energy extraction, the method that is used is greatly influenced by the quality of the specific source (Davis & Michaelides, 2009). Through heat exchangers, low-enthalpy geothermal energy can be utilized for direct heating. This is accomplished by rotating the geofluid in order to release its sensible heat. High-enthalpy geothermal sources can be used to generate electricity through steam power cycles, such as single, double flash, or dry steam. These cycles can be used to generate electricity (Alam, 2021). An ORC can be utilized as the primary source of energy for the generation of electricity from geothermal sources that have medium and low enthalpy. For example, hydrothermal regions that reach temperatures of up to 150°C can be utilized. This approach brings the heat from geothermal sources into the system in a roundabout way. In the ORC, continuous circulation of two fluids occurs: the geofluid, which is pumped from the low-enthalpy geothermal field and acts as the main fluid, and the organic working fluid, which is the secondary fluid and experiences thermodynamic changes (Sircar et al., 2007).

A number of countries, including New Zealand, Italy, Iceland, Kenya, the Philippines, and Mexico, are among the most significant countries engaged in geothermal energy projects. The temperature of geothermal water in Iceland can fluctuate anywhere from 200°C to 350°C. The area has fumaroles and hot springs that are heated by heat (Lu, 2018). There are more types of outdoor activities in places where the temperature is higher than in places where the temperature is low. In Iceland, fumaroles and mud pots, geysers, and hot springs have been found. More than 80% of Iceland's power comes from geothermal energy. The geothermal water in Kenya is 250–300°C and is about 2100 m below the surface (Yadav & Sircar, 2021).

There is a considerably lower enthalpy associated with India's geothermal energy potential in comparison to those of other countries. The So-Na-Ta Lineament, the Thermal

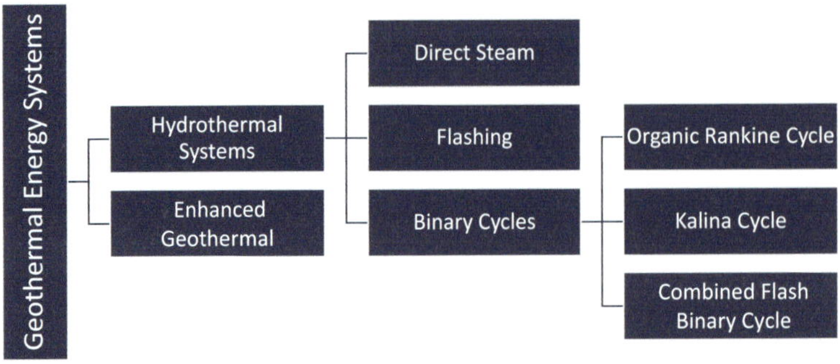

FIGURE 3.1 Classification of geothermal energy.

Field in the East and Northeast, the Northwestern Himalayas, the West Coast Margin, and Other Geothermal Regions are the five categories that India has placed its geothermal zones under by categorizing them. India is home to geothermal reservoirs that have temperatures that range from 30°C to 120°C (Singh et al., 2016). The enthalpy of the geothermal water in Puga is the highest among all other fields. Geothermal resources can be accessed through hot springs (Harinarayana et al., 2006). There has been limited exploration or exploitation in this industry due to its early stage of development. However, various organizations such as CEGE, NGRI, Thermax, and GSI are actively researching the potential applications of this energy (Bhatia, 2014; Chandrasekharam, 2005).

Figure 3.1 illustrates the classification of geothermal energy systems, which include hydrothermal and enhanced geothermal systems. They are further classified as direct steam, flashing, and binary cycles. In addition, the binary cycle subcategories are the ORC, Kalina cycle, and combined flash binary cycle.

As this chapter acknowledges the vast range of applications for geothermal sources, its primary focus will be on the viability of desalination processes powered by geothermal energy for producing freshwater. In the following sections, we will cover the rationale behind geothermal desalination, its important aspects, research on its practicality, case studies, and the global potential for its implementation in the near future.

3.3 INTEGRATED GEOTHERMAL DESALINATION SYSTEM

Desalination using geothermal water is a novel and cutting-edge idea (California et al., 2008). The use of geothermal water to desalinate saltwater has been the subject of research. Desalination makes use of low-temperature geothermal resources, while power generation makes use of high-temperature resources. An overview of geothermal-powered membrane, thermal, and hybrid desalination systems is given in this section (Jha et al., 2023).

It is possible to make use of the enthalpy of the geofluid in order to perform direct heating of the saltwater in distillation systems. In addition, the enthalpy of the geofluid can be exploited in an indirect manner to create electricity for the operation of RO units in integrated geothermal desalination (Prajapati et al., 2023). The primary heat exchanger, also known as the HE, is the component that is accountable for driving a desalination evaporator that operates at high temperatures and uses geothermal energy (Farsi & Rosen, 2022).

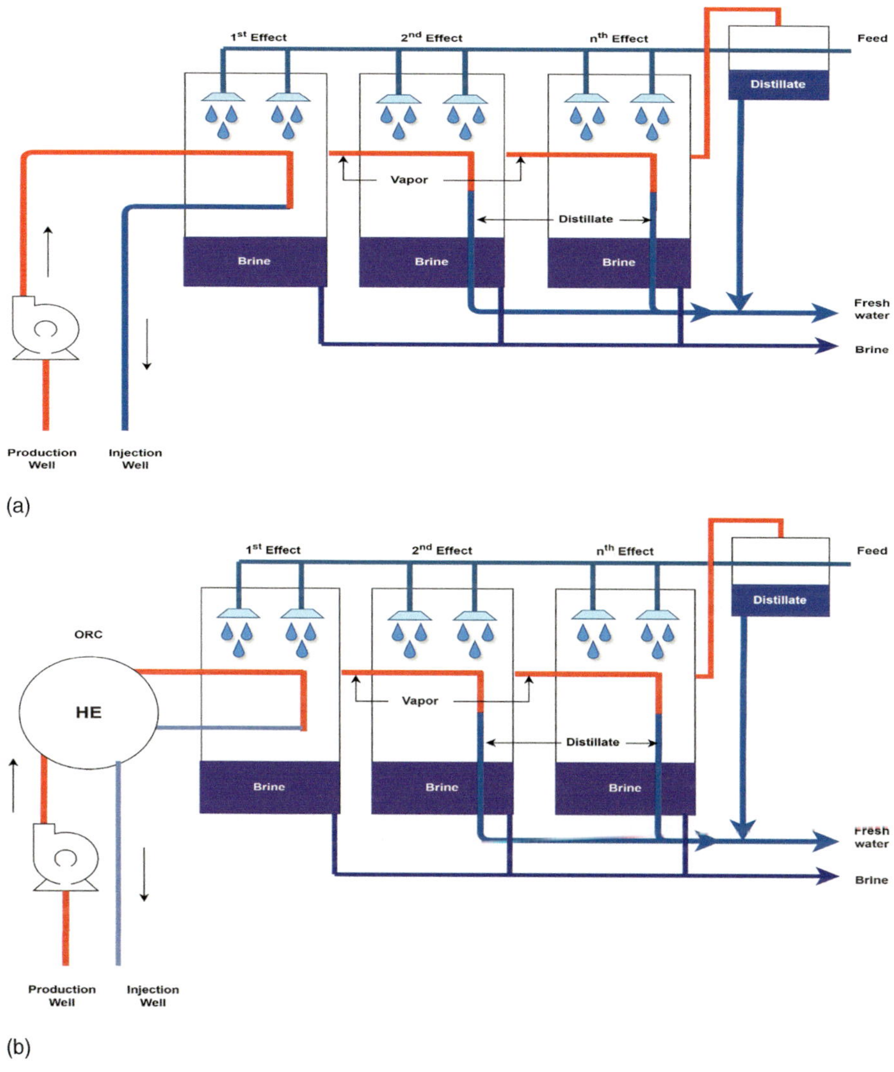

FIGURE 3.2 Utilization of geothermal heat for desalination (a) direct utilization (b) indirect utilization.

Figure 3.2 illustrates the utilization of geothermal heat for desalination. Figure 3.2 (a) shows the direct utilization of geothermal heat in the desalination process, and Figure 3.2 (b) shows the indirect utilization of geothermal heat in the desalination process.

Geothermal energy is a technology that has been demonstrated to be effective and is widely used for a variety of applications, including the generation of electricity, the cooling and heating of entire districts, and industrial processes (Gude, 2018). Thermal energy from geothermal sources is highly suitable for desalination due to its many benefits, including:

i Producing electricity and water simultaneously is feasible with geothermal desalination, making it a cost-effective option (Gude, 2016).
ii Thermal desalination (MED, MSF, etc.) and membrane-based (RO, etc.) processes rely on a steady supply of heat from geothermal energy sources because of their high-capacity factor, which is the availability of the resource in terms of quality as well as quantity (during the course of an application period).
iii The average temperatures of geothermal resources in most regions of the world typically fall within the range of 70–130°C, making them well-suited for low-temperature MED desalination. Sources with temperatures exceeding 100°C can produce electricity and other purposes requiring heat.
iv The extraction of hot water, also known as geofluid, from below-ground aquifers is not impacted by seasonal variations and weather changes.
v In contrast to other desalination processes, geothermal desalination does not release harmful air pollutants or greenhouse gases into the atmosphere because it relies solely on renewable energy sources. By saving imported fossil fuels, geothermal desalination helps to improve local energy security and sustainable development utilizing additional applications.

Geothermal desalination has been successfully implemented in many places within the literature. In 1959, Bell performed an economic evaluation of a multi-effect-evaporation plant that utilized geothermal energy as a heating source, with a daily capacity of 75 m^3 (Bell, 1959). After ten years, the US Department of Reclamation in California tried to produce freshwater directly from geothermal fluid at temperatures above 150°C. Various desalination systems, such as a vertical tube evaporator (VTE) and a multi-stage flash (MSF) pilot plant, have been assessed (Bermudez-Contreras et al., 2008). The initial patent for a desalination device specifically intended for high-enthalpy geothermal settings was granted in 1976 (Darwish et al., 1976). In 1982, Ophir conducted an initial economic analysis of geothermal desalination in Israel, focusing on temperatures ranging from 100°C to 150°C. In the year 1996, a geothermal desalination system known as Aero-Evapo-Condensation was put into operation in the southern region of Tunisia. It fed brackish geothermal water at a temperature of 65°C and had a capacity of 3 m^3 per hour. Within the framework of the approach, it was conceivable to make use of solar and/or geothermal energy at a cost that was fair while still retaining a straightforward process (Ophir, 1982; Ophir & Lokiec, 2005).

An innovative project was made by Karytsas et al. (2004) for the island of Milos, which is located in Greece, later in the year 2004. For the purpose of delivering freshwater through desalination and generating power through the utilization of low-enthalpy geofluid as the sole source of energy, they presented a hybrid system that was constructed and operated with the objective of accomplishing both things. Among the components that made up the hybrid system were the low-enthalpy MED unit, which had a capacity of 80 m^3 per hour, and the 470 kWe ORC power generation unit, which had a thermal efficiency of roughly 7%. Both of these units were able to generate power. Both of these components were fueled by the same geothermal resource that had a low thermodynamic energy. One of the most significant benefits of this system was that it was capable of providing its own thermal energy supply and the capacity to provide its own electrical energy supply (which was necessary for meeting the pumping requirements of the geothermal wells and the MED). This project's unit cost of water was estimated to be around $2 per m^3.

In 2008, an experiment was carried out in Baja California, Mexico, with the purpose of evaluating the potential for geothermal desalination in a variety of locations. Combining the multi-flash with heaters (MFWH) and multi-flash with heaters (MED) technologies resulted in the creation of a one-of-a-kind desalination device (Bermudez-Contreras et al., 2008). According to Abdelkareem et al. (2018), in order to create desalinated water, low-enthalpy geothermal energy was paired with two different geothermal desalination processes. These procedures were RO desalination and geothermal MED. The geothermal RO and MED systems are able to generate a daily output of 30,000 m^3 of drinkable water by utilizing the thermal energy that is generated by a geothermal dual power plant. In this particular scenario, a hybrid RO and MED unit that is powered by geothermal energy is able to generate freshwater with greater efficiency than a desalination plant that takes traditional energy sources into consideration. Both of these approaches have the potential to be very efficient and cost-effective technologies for desalinating saltwater in the coastal regions of the Gulf Cooperation Council states (Bachellerie, 2012).

A company in the United States carried out a number of studies during the course of a pilot project that utilized geothermal energy. These experiments reproduced the treatment variables in a VTE system that had a 15.0 effect. Using a VTE distillation process that is integrated in the design of a MED plant and using low-cost geothermal heat that is not commercially available as the energy source, it is possible that this technology will be more cost-effective than standard desalination techniques. One pound of geothermal steam that is utilized by a system with a 15.0 impact has the ability to generate up to 14 pounds of freshwater. This is the potential that exists. Utilizing this method, it is frequently possible to recover 80% or more from the source of the saltwater. With MED's assistance, the Salton Sea's salinity was decreased (Bullock & Gude, 2023). This was achieved by employing a VTE distillation method for geothermal desalination, powered by electricity produced by a geothermal power station. Consumption rates for the prototype and demonstration systems were 454 kg per hour and 79.50 m^3 per day, respectively. In a study conducted on the Greek island of Milos (Karvounis, 2017), Karystsas presents the results of an investigation into a desalination system that was operated using geothermal energy with a reduced enthalpy. The suggested system connects MEDs to a geothermal groundwater source with a temperature between 80°C and 85°C (S. Karytsas et al., 2019).

Missimer et al. (2016) suggested a prototype plant that combines geothermal power generation with desalination to produce electricity and freshwater simultaneously. The system extracts extremely hot steam, around 300°C and 6-7 MPa in pressure, from the producing well. After that, this steam is utilized to operate the steam turbine within the system. Electricity produced by the steam turbine is used to operate a RO desalination system. In addition, the turbine outlet steam is utilized to provide power to MED and membrane desalination facilities. In their study, Petrovic et al. (2018) examined the thermodynamics of a MED desalination facility connected to hot springs and utilized ejector refrigeration. An optimization analysis was conducted to minimize the overall investment costs and exergy loss rate while maximizing the facility's profits. Furthermore, a comparison was made between different refrigerants in terms of their impact on the performance and cost of the refrigeration system.

Gude (2018) examined the utilization of geothermal energy specifically for generating drinkable water. The paper explains both the benefits and drawbacks of geothermal energy. Geothermal energy can be utilized through MSF distillation or MED to generate power and produce potable water simultaneously. Furthermore, plant-based geothermal energy is built using inexpensive materials, resulting in reduced rates of corrosion, scaling, and heat losses. The start-up durations are shorter, and the plant's lifespan is extended.

A feasibility study was conducted in Baja California, Mexico, to determine whether or not the utilization of geothermal sources would be feasible. A multi-effect boiling unit and an MSF unit were both parts of the integrated design that was put through various tests in this study. A temperature of 80°C was available from the geothermal source, which could be used as a source of heat. According to the results of this technique, the optimal proportion of desalinated water (freshwater) to geothermal supplies (heat source water) is 13 to 1. Using the MSF-MED layout, a coastal geothermal desalination plant was investigated as part of a subsequent research effort. At a temperature of 80°C, the prototype was designed and tested when it was being reviewed. The ratio of 1:5.9 between the freshwater and geothermal sources was found to be much lower, according to the findings of this inquiry. The creation of around 20 m^3 of freshwater on a daily basis was the outcome of a geothermal heat flow of 118 m^3 per day (Bermudez-Contreras et al., 2008).

According to Birney et al. (2019), the state of Texas has successfully built an MED plant that makes use of geothermal energy. This plant has achieved an average daily freshwater production rate of 120 m^3. The greatest temperature that can be found underground in Texas is roughly 150°C, and it may be found at depths ranging from 3000 to 4000 m. According to Birney et al. (2019), increasing the geothermal fluid mass flow rate from 30 to 90 kg/s results in a commensurate rise in the amount of clean water generated in the geothermal MED system, which goes from 121 to 1132 m^3 per day.

El Haj Assad et al. (2022) introduced a dry steam geothermal power plant that operated an RO saltwater desalination system. In addition, the dry steam system was employed to power the pumps of the RO unit. A single-stage, single-pass RO setup, connected to a pressure exchanger, was employed to desalinate seawater with a salinity of 48,787 ppm. The process consumed 3.3 kWh per m^3 of water, with a daily productivity of 450 m^3 and a product water salinity of 335.58 ppm. The geothermal energy system was specifically developed to generate 350,400 kWh/year of power. The optimization resulted in an average electrical output of 40 kW and an automatically adjusted backup generator rating of 98 kW.

An overview of the process conditions for different geothermal sites worldwide is included in Table 3.1 provides an overview of the various desalination methods and the temperatures and flow rates of the geothermal water needed for each. The techniques include membrane desalination and thermal desalination. Costs and the amount of geothermal water flow are affected by the settings of the desalination process.

3.4 CASE STUDIES

3.4.1 Geothermal Desalination at Baja Peninsula, Mexico

Over 1400 geothermal sites have been discovered in Mexico (V Torres-Rodriguez, 2000). Most of these geothermal sites are located inland and are primarily assigned to power production. This is because Mexico has installed more than 840 MW of geothermal capacity. On the other hand, the east Pacific Rise along the Baja Peninsula is home to several geothermal marine resources. These resources are located where continuous heat or hot water discharge occurs almost directly on the coast.

However, the California Peninsula is very separated from the rest of Mexico, and the supply of petroleum to this region is quite expensive. Despite this, the Baja California Peninsula, particularly the Los Cabos area (located in the state of Baja California Sur), is one of the most popular tourist destinations in Mexico (Bermudez-Contreras et al., 2008). There has been a significant rise in the demand for electricity and water as a direct result of the expansion of tourism, which has led to an increase in the population. A source of hot water, which was most likely associated with a hydrothermal system, occurred while drilling some wells to supply the desalination plant in Cabo San Lucas. As a result, geothermal energy has been viewed as a potential solution to the issues plaguing the energy and freshwater sectors. Not only does geothermal energy provide electricity, but the hot water that is produced can also be utilized in desalination operations. In addition, hot water can be utilized in the spa facilities, increasing the availability of tourist attractions in the area. We describe the findings of geochemical and isotopic analyses performed on three geothermal wells drilled in this region. In order to determine the origin of these waters and the processes that may have been responsible for their formation, as well as to estimate the temperature of the reservoir, these analyses were utilized (Saeed et al., 2019).

In the past ten years, one of the most popular tourist zones in Mexico, which is situated at the southernmost point of the Peninsula of Baja California, has experienced spectacular growth in the number of hotels and the infrastructure dedicated to providing services. Both the demand for water and the demand for power in the region have significantly increased as a consequence of this. An operator of the municipal water system

TABLE 3.1 A summary of geothermal desalination installations around the world

LOCATION	*YEAR*	*INPUT WATER*	*CAPACITY*	*DESALINATION METHOD*	*HEAT UTILIZATION*	*REMARK*	*OPERATIONAL COST (PER/M³)*	*REFERENCE*
Baja California, Mexico	2008	Seawater	1 m^3/day	MSF and MED	Geothermal energy	• The MFWH is a hybrid system that combines MED & MSF technologies. • When the temperature was first set at 150°C, it took 4 m^3 of seawater to generate 1 m^3 of desalinated water. A volume of 14 m^3 of geothermal water was needed at a starting temperature of 80°C.	-	
Kimolos, Greece	2002	Brackish water	80 m^3/day	MED	Geothermal energy	• A two-stage MED process incorporating vacuum distillation in vertical tubes. • The geothermal water is flowing at a rate of 60 m^3 per hour, with a wellhead temperature ranging from 61°C to 62°C. The water is coming from a borehole that is 188 m deep. • The daily generation of freshwater amounts to 80 m^3. The anticipated cost of produced water is approximately 1.6€ per m^3, considering only the annual running expenses.	EUR 1.60	

Tunisia, South Tunisia, France,	1996	Brackish water	3 m^3/h	MED	Low-enthalpy geothermal source	• MD coupled with multiple-effect - distiller. • The investigation determined that the optimal operating conditions consist of a feed brine temperature of 85°C at the evaporator intake and a circulation flow rate of approximately 170 kg/h. Given these circumstances, it is possible to achieve a GOR (Gas-Oil Ratio) value of 3.7 and produce 16 kg/h of water. By incorporating a membrane module distiller as a secondary step at the outlet of the MED process, the production of distilled water may be increased by around 7.5%, and the energy efficiency can be improved by nearly 10%.	
MilosIsland, Greek	2004	Seawater	80 m^3/h	MED	Solar energy and lower enthalpy geothermal	• The ORC unit generated 470 kW of power with a thermal efficiency of 7%. The cost per unit of producing freshwater is approximately \$2 per m^3.	EUR 1.50

(*Continued*)

TABLE 3.1 (Continued) A summary of geothermal desalination installations around the world

LOCATION	YEAR	INPUT WATER	CAPACITY	DESALINATION METHOD	HEAT UTILIZATION	REMARK	OPERATIONAL COST (PER/M³)	REFERENCE
Arabian Gulf Country	2013	Seawater	30,000 m³/day	MED and RO	Grid electricity, geothermal energy, PV	• The plant has a yearly availability - of 92% and an accessibility rate of 8% as a result of plant support and other contributing variables. Continuous operation for 24 hours every day.		
USA		Seawater	79.5 m³/day	ED/VTE	Geothermal energy	• The facility made use of a 15-effect evaporation unit. • The steam temperature is 100°C, and the flow rate is 3402 kg per hour.	$0.65	
USA		Seawater	18.9 m³/day	MED/VTE	Geothermal energy	• The temperature of the steam is 100°C at a flow rate of 454 kg per hour. This stat on employs a two-fold effect evaporation technology.	$0.65	

owns and operates a desalination plant that has a capacity of 200 L per second on a daily basis. Initially, they bore a few wells with temperatures ranging from 35°C to 72°C (López-Ramírez et al., 2005), one of which was 84°C. However, they could not use the hot water they drilled in the RO process.

In order to make use of the geothermal resources (98°C) that have been discovered on the beach there, IMPULSA established two small-scale prototype plants in La Joya, a leisure camp located south of the city of Ensenada in British Columbia. This would be performed as a demonstration project. When a geothermal shallow well of 50 m in depth is drilled close to the cost, it will generate sufficient heat output (140°C) to power a small power generator. This electricity could then be used to install and operate RO technology, which would be used to desalinate seawater. The primary objective is to examine the response of the geothermal reservoir at La Joya, while simultaneously testing and evaluating the worldwide performance of both plants for one year, respectively. Utilizing a combination of a MED and an MSF thermal desalination plant, the LE-MED system is an innovative IMPULSA design developed from the two technological combinations. In classic MSF and MED facilities, the process begins with preheating seawater to its boiling point through steam, which is often generated from a thermal power generator. Following this, the seawater that has been warmed is evaporated, and the steam that is free of salts is condensed as freshwater. The plant is discharged with the remaining water as concentrated brine. The new IMPULSA LE-MED does not use steam in any way during the process; rather, geothermal hot water is utilized as a source of energy to support the system's operation. Within the desalination process framework, this LE-MED system's primary objective is to keep clear of the utilization of sources derived from fossil fuels. The price of desalinated cubic meters would be greatly reduced due to these (Gutiérrez et al., 2010).

Figure 3.3 illustrates a detailed map of Mexico highlighting the location of the Geothermal desalination at Milos Island, Greece.

Milos is part of the active volcanic arc in the South Aegean region. The island has a strong geothermal gradient because of the recent volcanic activity that occurred 80,000 years ago and molten magma at a depth of a few kilometers. The intense tectonic activity and frequent earthquakes on Milos contribute to the ideal circumstances for the geothermal gradient. This has been confirmed by the drilling of five deep wells (ranging from 1100 to 1350 m) in the middle eastern region of the island (Fytikas et al., 2005b).

In Greece, the eastern side of the island, and more specifically the plain of Zefyria, displayed the most significant temperature changes (C. Karytsas et al., 2004). As a result, this region was the most suitable location for the development of high-enthalpy geothermal resources. The further drilling exploration that was conducted indicated the presence of geothermal fluids at depths ranging from 800 to 1400 m below sea level. These fluids had temperatures that ranged from 300°C to 323°C, respectively. S. Karytsas et al. (2019), on the other hand, found that the eastern half of the island was the most favorable region for using shallow geothermal resources that had a low enthalpy (less than 100°C). The deep geothermal fluids consisted of saltwater that had been heated to the boiling point and had a salinity of 80,000 parts per million. It was concluded that the upper 2 km of the heated rocks beneath Zefyria can sustain a geothermal power station with a power output of 260 MWe.

FIGURE 3.3 Geothermal desalination at Milos Island, Greece (Fytikas et al., 2005a).

According to C. Karytsas et al. (2004), the main goal of the Milos Island project was to construct and run a desalination plant that used low-enthalpy geothermal energy to create 80 m^3/h of potable water and a 470 kWe power-generating unit. This was the project's principal aim. There were two systems at work at the plant: first, a multiple-effect evaporator or distillation unit (MED) used the hot water directly; second, ORC turbines used the water to generate electricity. These two systems were used in tandem. A working fluid is pumped to a boiler, evaporates, and then re-condenses in the ORC, which is conceptually similar to the Rankine cycle. The fluid then flows through a turbine to produce energy.

Also, multiple-effect distillation (MED), a boiling method, is often used in operations to remove salt from seawater. This has several steps or "effects" that make it up. The steam inside the tubes heats the feed water at every level. Some of the pure water disappears, and the steam is then sent to the tubes of the next stage, where it heats up and evaporates more water. In essence, each successive stage recycles the energy used in the one before it. It is possible to submerge the tubes in the feed water; however, the more common practice is to spray the feed water, which is colder and saltier, on the top of a bank of horizontal hot tubes that hold the pure water vapor. Condensation occurs as the vapor in the tube's drips from one tube to the next until it is ultimately collected at the base of the stage. The Milos Island initiative is expected to be self-sufficient in both thermal and electrical energy, as geothermal heat is the primary energy source that will

FIGURE 3.4 Vounalia area of Milos island (Fytikas et al., 2005a).

be employed. The local community would benefit from the use of a low-enthalpy geothermal energy source that is both sustainable and environmentally benign. Furthermore, the local populace would derive advantages from an uninterrupted supply of desalinated water at a reasonable cost.

Figure 3.4 illustrates a detailed map highlighting the location of the Vounalia area of Milos Island.

3.4.2 Case Study of Geothermal Desalination Potential in India

A variety of significant geothermal energy sources may be found in India. Some of these are the Puga Valley in Jammu and Kashmir, Manikaran in Himachal Pradesh, Chhumathang in Jammu and Kashmir, Bakreshwar in West Bengal, and Unai in Maharashtra (Puppala & Jha, 2021). Additionally, a new source of geothermal electricity has been found in the state of Chhattisgarh, which is situated on the Narmada River in the central region of India (Singh et al., 2018). In addition, Gujarat is launching a project to create power by making use of the geothermal energy that can be discovered in the Cambay region, which is situated between the Narmada and Tapi rivers (Puppala & Jha, 2018). Geothermal energy in India has primarily been used in pilot plants for scientific and technological development purposes up until this point. Under the guidance of the National Aeronautical Laboratory (NAL) and the Geothermal Science Institute (GSI), an experimental geothermal energy power plant with a capacity of 5 kW has been constructed in the vicinity of Manikaran. This operation uses a closed-loop Rankine cycle, which is carried out with the Freon-113 working fluid. According to Ghose (2004), India has not yet built larger geothermal power facilities, despite NAL's involvement in their planning and manufacturing. Around the world, governments are generally positive about the idea that opening up the economy could lead to more private investment in geothermal power creation. Pillai and Banerjee (2009) say that geothermal power plants are already up and running in a number of countries. India has made a big plan to increase its production capacity from 100,000 MWe to 215,000 MWe. If it succeeds, it will no longer need power. According to Singh et al. (2016), it is anticipated that carbon-based, petroleum-based, nuclear, and substantial hydroelectric projects will be able to accomplish this objective (Figure 3.5).

In Gujarat's broad spectrum of lithology, there are 22 hot springs with temperatures ranging from 35°C to 93°C. The Cambay Basin is the most prominent tectonic structure in the state. A historic city in Dholera is about 60 km away from the large city of Bhavnagar (Shah, Sircar, et al., 2018a). Hot springs known as Dholera can be found in this region, specifically along the west seashore of the Saurashtra Peninsula, which includes the West Marginal fault of the Cambay Basin. Sircar et al. (2015) reported that the sediment cover in Dholera is approximately within the range of 500.0–600.0 m. Several shallow bore wells have been drilled in Dholera and nearby communities for the purpose of irrigation and household use. The geothermal water temperature in this region, as reported by Shah, Vaidya, and Sircar (2018b), falls within the range of 40.0°C to 45.0°C.

Chandrasekharam et al. (2016) conducted a study focusing on provinces with geothermal energy. One of the most promising provinces in the very cold section of the country, the Himalayan Province, is home to almost ten thousand hot springs with subsurface temperatures exceeding 90°C. The reservoir of the tank has a temperature of around 260.0°C. Intruding post-tertiary granite is the cause of the temperature gradient of 100.0°C per km and the heat flux of 468.0 MW/m^2. More than 50,000 m of the post-Cretaceous sedimentary deposit is a particularly attractive site for geothermal energy. This deposit is located above the flood basalts of the Deccan Plateau. More than 15 thermal discharge spots can be found in this region, and temperatures near the surface at these locations range from 40°C to 90°C. In a reservoir with a heat flow of 80–93 MW/m^2, the temperature ranges from 150.0°C to 175.0°C.

In Dholera, Gujarat, the unit is intended to desalinate the water produced by geothermal sources. The geothermal water in the area is quite shallow, and as a result, it is found as groundwater when it is brought up for drinking reasons. Python programming is used to design an MSF distillation system. The selection of the MSF system was made to produce freshwater of superior quality while reducing the amount of salt present in the water. This plant has dual applicability: first, it may be utilized for freshwater production, and, second, it can be employed for salt production from the discharged brine. Furthermore, the aim was to convert the whole system into an ecologically sustainable and renewable one; consequently, the solar evacuated tubes function as an energy source. The selection of solar collectors was dependent on the minimum temperature requirements of the MSF system. The proposed system has the capability to not only

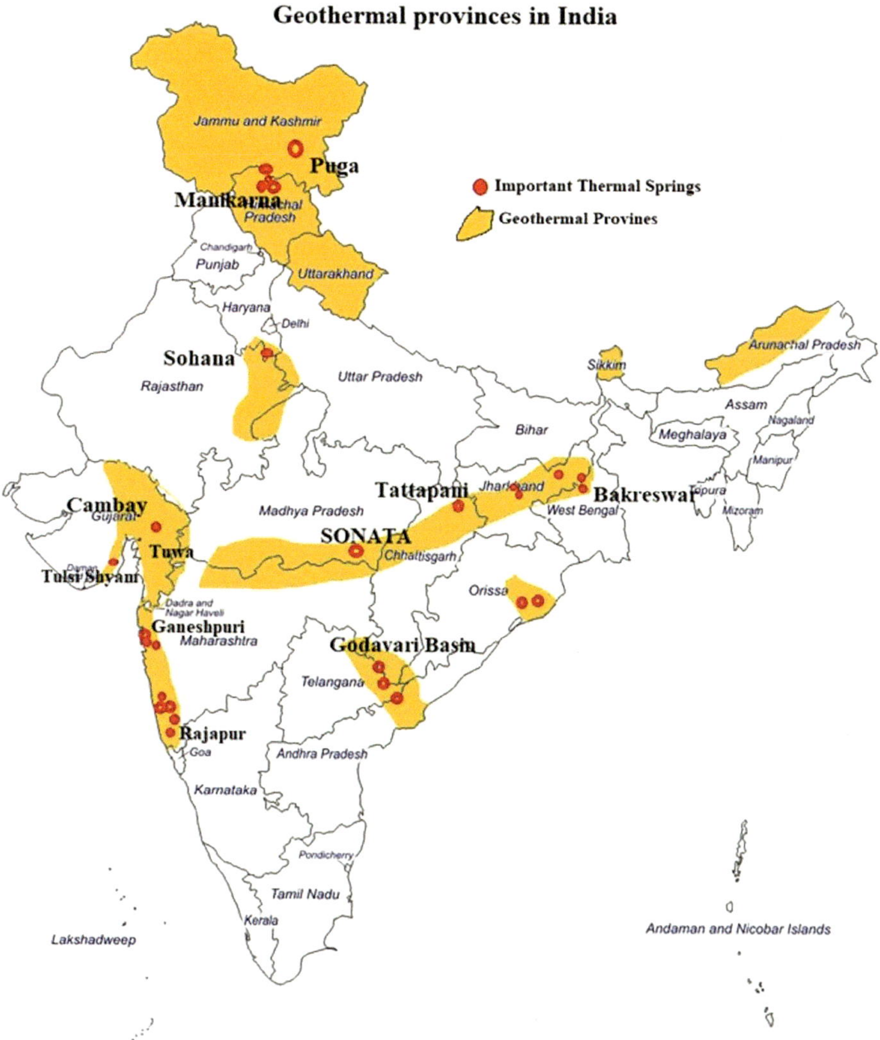

FIGURE 3.5 Geothermal sites in India (Prajapati et al., 2022).

produce salt but also meet the drinking water requirements of the neighboring residents. The heat rejected by the system can also be utilized for other reasons, such as drying food, thermal therapy, and other similar activities. The simulation yielded highly satisfactory results, as the salt concentration of the water was successfully reduced to the standards of freshwater, namely from a maximum of 194 parts per million (PPM). Furthermore, the system was cost-estimated by considering both capital and operational expenditures. According to estimates, the cost of water per cubic meter was \$0.1034 (Yadav et al., 2023).

Figure 3.5 illustrates a map of India highlighting the locations of various geothermal sites across the country.

3.4.3 Geothermal Desalination in the United States

Desalination and power generation through geothermal energy sources near the Salton Sea in California. Geographically, the Salton Sea is situated in the heart of a region rich in agricultural output. The Imperial County Farm Bureau (ICFB) reports that agriculture consumes an average of 5.6 acre-feet of water, per acre, annually. This information is according to the ICFB. One ton of salt is transported into the region for every acre-foot of water brought into the Valley through the drainage system. Five hundred and twenty trucks would be required to transport all of the salt dumped into the Salton Sea annually. All runoff waters from the neighboring areas are directed toward the Salton Sea by the drainage systems used in those places. The sea and the places surrounding it have experienced an exponential salinity increase due to this matter (Bullock & Gude, 2023).

Further investigation has revealed that the sea is actively accessing nearby geothermal reservoirs, resulting in an elevation of the salinity of the geofluid. This phenomenon is ongoing as the salt concentrations in the water continuously increase. Both brine from geothermal reservoirs and brine from the Salton Sea are appropriate options for the desalination process since they have total dissolved solid concentrations ranging from 20% to 25%. According to the United States Bureau of Reclamation (USBOR), the total dissolved concentrations in the water of the Salton Sea are estimated to be around 44,000 mg/L. Nevertheless, there have been earlier reports indicating a higher concentration of around 55,000 mg/L. It is interesting to note that it is dominated by salt chloride concentrations comparable to those found in geothermal brines and groundwater in the district (Bullock & Gude, 2023).

There is a possibility that the villages surrounding the Salton Sea could utilize geothermal sources in order to cure the salty groundwater supplies. The elevated temperatures of the geofluid indicate potential for desalination use. The benefits of the desalination plant might be utilized by the local towns and enterprises already in operation to fulfill the requirements for future water and energy requirements. Although the geothermal area is situated at a considerable distance from the ocean, the water quality is considered sufficient for the treatment process.

Two case studies were conducted as part of the investigation. One case study examined the energy requirements of a desalination plant considering both thermal and electrical energy. The second case study examined the same plant but only considered electrical energy, assuming that the thermal energy wasn't a cost because it was directly derived from the geothermal water from the wells. The cost per megawatt of energy for three different desalination processes—MED, RO, and MSF distillation—was meticulously examined in each of the studies. Various energy levels were used to conduct the analysis.

The analysis indicates optimal results for establishing a geothermal desalination plant in the Salton Sea Area. This is the case when the water and power strategy is considered. The energy produced throughout the day is a reasonable explanation for the expense of the plant, as the net nominal expenses are approximately 16 cents/kWh on average. If the facility's total capacity is increased, the net installation cost, also known as capital expenditure, will grow. However, the difference between the two is exactly what was anticipated. A fair comparison may be made between the desalination power plant that operates independently and the desalination power plant that generates electricity (Alhuyi Nazari et al., 2023).

Some areas in the Middle East have been discovered to have significant geothermal energy. Take, for instance, the country of Saudi Arabia, which is home to some significant geothermal hotspots (Hossam-Eldin et al., 2017). Geothermal sources in Saudi Arabia can be classified into three main categories: low-enthalpy resources, predominantly located in the eastern region; medium-enthalpy resources, distributed in the western and southwestern regions; and high-enthalpy resources, composed of basaltic lavas and Harrats. Within Saudi Arabia, there are two specific areas that possess a high concentration of geothermal resources. Although sedimentary basins in the Arabian Gulf offer significant low-enthalpy resources suitable for desalination, cooling, and low-temperature steam production, volcanic communities in the Red Sea can potentially provide medium to high-enthalpy resources for power generation (Alharbi et al., 2012; Nexus et al., 2018).

Nisyros, a Greek island that is located in the southeastern part of the Aegean Sea was the subject of a recent research that investigated the geothermal potential of the country. The studies focused on seawater desalination and power production and were conducted using independent and coupled schemes. In this particular investigation, an MED was selected as the method of choice since it has reduced specific energy requirements and smaller scaling potentials due to its application at low temperatures. In addition, an examination of the desalination plant's energy consumption was carried out in order to gain a better understanding of the plant's thermodynamic performance. In addition, a life cycle analysis was also carried out using the desalination power production coupling method MED utilized. It was reported that the environmental implications were insignificant compared to a cogeneration plant fueled by fossil fuels. The desalination process also made use of the waste heat from the power plant, which resulted in the elimination of the environmental emissions that were associated with the process (Karvounis, 2017).

3.5 CONCLUSION

In order to combat freshwater scarcity, this chapter primarily discusses methods that combine geothermal energy with desalination procedures. This approach is especially useful in areas where water resources are scarce. It emphasizes how pollution, climate change, and increasing human consumption are increasing the need for freshwater. Using geothermal energy as an example, this chapter analyzes its desalination potential in depth and stresses the significance of moving away from fossil fuels and toward renewable energy sources. Geothermal energy, characterized by its temperature and enthalpy levels, is presented as a sustainable and efficient solution for producing freshwater. The chapter discusses various geothermal desalination systems, their operational principles, and case studies from around the world, showcasing successful implementations and the potential for future expansion.

In conclusion, integrating geothermal energy into desalination processes presents a promising solution to address the global water scarcity challenge. The utilization of geothermal energy, a renewable and stable heat source, for desalination has been explored through various methods such as MSF, MED, and RO, powered by both high- and low-enthalpy geothermal sources. Case studies from Mexico, Greece, Tunisia, and the United States demonstrate the feasibility and efficiency of geothermal desalination systems, highlighting their potential to produce freshwater at a lower environmental impact and operational cost than conventional desalination methods. This approach offers a sustainable alternative for freshwater production, contributes to local energy security, and supports sustainable development by reducing reliance on imported fossil fuels. As the demand for freshwater rises, geothermal desalination stands out as a viable and eco-friendly option that can be adapted to various geographical contexts and energy profiles.

REFERENCES

Abdelkareem, M. A., El Haj Assad, M., Sayed, E. T., & Soudan, B. (2018). Recent progress in the use of renewable energy sources to power water desalination plants. *Desalination*, *435*, 97–113. https://doi.org/10.1016/j.desal.2017.11.018

Alam, M. A. (2021). *Tectonic and Structural Controls on Geothermal Systems*. 223–242. https://doi.org/10.1007/978-981-16-4122-0_15

Alharbi, O. A., Phillips, M. R., Williams, A. T., Gheith, A. M., Bantan, R. A., & Rasul, N. M. (2012). Desalination impacts on the coastal environment: Ash Shuqayq, Saudi Arabia. *Science of the Total Environment*, *421–422*, 163–172. https://doi.org/10.1016/j.scitotenv.2012.01.050

Alhuyi Nazari, M., Kumar, R., Mukhtar, A., Yasir, A. S. H. M., Ahmadi, M. H., & Al-Bahrani, M. (2023). Geothermal energy for preheating applications: A comprehensive review. *Journal of Central South University*, *30*(11), 3519–3537. https://doi.org/10.1007/s11771-023-5410-9

Almtairi, A., Sharaf Eldean, M. A., Soliman, A. M., Mabrouk, A., & Fath, H. E. S. (2021). A new preliminary system design of using geothermal well brine heater for desalination/nanofiltration process. *Cleaner Engineering and Technology*, *4*(June), 100213. https://doi.org/10.1016/j.clet.2021.100213

Aviña-Jiménez, H., De La Rosa-Urbalejo, D., & García-Gutiérrez, A. (2016). Desalination using geothermal energy. *Transactions - Geothermal Resources Council*, *40*, 141–145.

Bachellerie, I. J. (2012). Renewable Energy in the GCC Countries. In *Gulf Research Center*. Gulf Research Center.

Bell, J. (1959). *Availability of geothermal energy for the demineralization of saline water*. https://books.google.com/books?hl=en&lr=&id=DLiMmuFHTOgC&oi=fnd&pg=PP9&dq=Availability+of+Geothermal+Energy+for+the+Demineralization+of+Saline+Wate&ots=Nl5J4KpRxJ&sig=0T8Qyvh8J8uacT5bIJDqAnuaLQA

Bermudez-Contreras, A., Thomson, M., & Infield, D. G. (2008). Renewable energy powered desalination in Baja California Sur, Mexico. *Desalination*, *220*(1–3), 431–440. https://doi.org/10.1016/j.desal.2007.01.046

Bhatia, S. C. (2014). Geothermal energy resources and its potential in India. *Advanced Renewable Energy Systems*, *I*(I), 389–402. https://doi.org/10.1016/b978-1-78242-269-3.50015-2

Birney, C. I., Jones, M. C., & Webber, M. E. (2019). A spatially resolved thermodynamic assessment of geothermal powered multi-effect brackish water distillation in Texas. *Resources*, *8*(2), 1–20. https://doi.org/10.3390/resources8020065

Bora, M. C. (2016). *Geothermal Energy: Indian Scenario Geothermal Energy: Indian Scenario. January*. https://doi.org/10.13140/2.1.2508.8969

Bullock, J., & Gude, V. G. (2023). Economics of brine desalination for communities near the Salton Sea Geothermal Field, California, USA. *Environmental Earth Sciences*, *82*(1). https://doi.org/10.1007/S12665-022-10649-6

California, I. B., Hernandez, S. E., & Ingenieria, T. De. (2008). *Using Coastline Geothermal Resources for Desalination. October*, 1–5.

Chandrasekharam, D. (2005). Geothermal Energy Resources of India - Past and the Present. *Proceedings World Geothermal Congress, April*, 24–29.

Chandrasekharam, D., & Bundschuh, J. (2008). Low-enthalpy geothermal resources for power generation. In *Low-Enthalpy Geothermal Resources for Power Generation*. https://doi.org/10.1080/00207233.2010.498602

Chandrasekharam, D., Lashin, A., Al Arifi, N., & Al-Bassam, A. M. (2016). Red sea geothermal provinces. In *Red Sea Geothermal Provinces*. https://doi.org/10.1201/b20408

Damania, R., Desbureaux, S., Hyland, M., Islam, A., Moore, S., Rodella, A.-S., Russ, J., & Zaveri, E. (2017). Uncharted Waters: The New Economics of Water Scarcity and Variability. In *Uncharted Waters: The New Economics of Water Scarcity and Variability*. https://doi.org/10.1596/978-1-4648-1179-1

Darre, N. C., & Toor, G. S. (2018). Desalination of Water: a Review. *Current Pollution Reports*, *4*(2), 104–111. https://doi.org/10.1007/s40726-018-0085-9

Darwish, M. A., Serageldin, S. G., & Eldessouky, H. T. (1976). Flashing in desalination units. *Desalination*, *19*(1–3), 93–101. https://doi.org/10.1016/S0011-9164(00)88020-0

Davis, A. P., & Michaelides, E. E. (2009). Geothermal power production from abandoned oil wells. *Energy*, *34*(7), 866–872. https://doi.org/10.1016/j.energy.2009.03.017

Dawoud, M. A. (2012). Environmental Impacts of Seawater Desalination: Arabian Gulf Case Study. *International Journal of Environment and Sustainability*, *1*(3). https://doi.org/10.24102/ijes.v1i3.96

DiPippo, R. (2016). Geothermal Power Generation: Developments and Innovation. In *Geothermal Power Generation: Developments and Innovation*. https://doi.org/10.1016/C2014-0-03384-9

El Haj Assad, M., Nooman Almallahi, M., Abdelsalam, M. A., Alshabi, M., & Almallahi, W. N. (2022). Desalination Technologies: Overview. *2022 Advances in Science and Engineering Technology International Conferences, ASET 2022*. https://doi.org/10.1109/ASET53988.2022.9734991

El-Ghonemy, A. M. K. (2012). Future sustainable water desalination technologies for the Saudi Arabia: A review. *Renewable and Sustainable Energy Reviews*, *16*(9), 6566–6597. https://doi.org/10.1016/j.rser.2012.07.026

Farsi, A., & Rosen, M. A. (2022). Assessment of a Geothermal Combined System with an Organic Rankine Cycle and Multi-effect Distillation Desalination. *Earth Systems and Environment*, *6*(1), 15–27. https://doi.org/10.1007/s41748-021-00275-w

Fewtrell, L. (2013). Water Quality: Guidelines, Standards and Health: Assessment of Risk and Risk Management for Water-Related Infectious Disease. *Water Intelligence Online*, *12*, 1–431. https://doi.org/10.2166/9781780405889

Fytikas, M., Radoglou, G., Karytsas, C., Mendrinos, D., Vasalakis, A., & Andritsos, N. (2005a). Geothermal Research in Vounalia Area, Milos Island, for Seawater Desalination and Power Production. *Proceedings World Geothermal Congress*, *2*, 24–29.

Fytikas, M., Radoglou, G., Karytsas, C., Mendrinos, D., Vasalakis, A., & Andritsos, N. (2005b). Geothermal Research in Vounalia Area, Milos Island, for Seawater Desalination and Power Production. *Proceedings World Geothermal Congress*, *2*, 24–29.

Ghaffour, N., Lattemann, S., Missimer, T., Choon Ng, K., & Amy, G. (2013). Seawater desalination using renewable energy: solar, geothermal, and wind 1 2. In *Proceeding of International Conference on Applied Energy (ICAE 2013)*.

Ghose, M. K. (2004). Environmentally sustainable supplies of energy with specific reference to geothermal energy. *Energy Sources*, *26*(6), 531–539. https://doi.org/10.1080/00908310490427977

Goosen, M., Mahmoudi, H., & Ghaffour, N. (2010). Water Desalination using geothermal energy. *Energies*, *3*(8), 1423–1442. https://doi.org/10.3390/en3081423

Green, S., McLennan, J., Panja, P., Kitz, K., Allis, R., & Moore, J. (2021). Geothermal battery energy storage. *Renewable Energy*, *164*, 777–790. https://doi.org/10.1016/j.renene.2020.09.083

Gude, V. G. (2016). Geothermal source potential for water desalination - Current status and future perspective. *Renewable and Sustainable Energy Reviews*, *57*, 1038–1065. https://doi.org/10.1016/j.rser.2015.12.186

Gude, V. G. (2018). Geothermal Source for Water Desalination-Challenges and Opportunities. In *Renewable Energy Powered Desalination Handbook: Application and Thermodynamics* (Vol. *1*). Elsevier Inc. https://doi.org/10.1016/B978-0-12-815244-7.00004-0

Gutiérrez, H., Bulletin, S. E.-G. et al. (2010). Using low enthalpy geothermal resources to desalinate sea water and electricity production on desert areas in Mexico. *Geothermal-Energy. Org*, 25–29. https://www.geothermal-energy.org/pdf/IGAstandard/WGC/2010/2808.pdf

Harinarayana, T., Abdul Azeez, K. K., Murthy, D. N., Veeraswamy, K., Eknath Rao, S. P., Manoj, C., & Naganjaneyulu, K. (2006). Exploration of geothermal structure in Puga geothermal field, Ladakh Himalayas, India by magnetotelluric studies. *Journal of Applied Geophysics*, *58*(4 SPEC. ISS.), 280–295. https://doi.org/10.1016/j.jappgeo.2005.05.005

Hossam-Eldin, A. A., Youssef, K. H., & Kotb, H. (2017). Technical and economic optimization of reverse osmosis desalination systems integrated with PV/wind energy resources: A case study. *2017 19th International Middle-East Power Systems Conference, MEPCON 2017 - Proceedings*, *2018-Febru* (March 2018), 510–514. https://doi.org/10.1109/MEPCON.2017.8301228

Jha, P., Rajguru, S., Parmar, J., & Shah, M. (2023). A holistic study on geothermal and solar water desalination system for sustainable development. *Environmental Science and Pollution Research*, *30*(1), 90–103. https://doi.org/10.1007/s11356-022-22987-w

Karvounis, P. (2017). A Review of Desalination Potential in Greek Islands Using Renewable Energy Sources, a Life Cycle Assessment of Different Units. *European Journal of Sustainable Development*, *6*(2), 19–32. https://doi.org/10.14207/ejsd.2017.v6n2p19

Karytsas, C., Mendrinos, D., & Radoglou, G. (2004). The current geothermal exploration and development of the geothermal field of Milos island in Greece. *GHC Bulletin*, *25*(June), 17–21.

Karytsas, S., Polyzou, O., & Karytsas, C. (2019). Social aspects of geothermal energy in Greece. *Lecture Notes in Energy*, *67*, 123–144. https://doi.org/10.1007/978-3-319-78286-7_9

Koroneos, C., & Roumbas, G. (2012). Geothermal waters heat integration for the desalination of sea water. *Desalination and Water Treatment*, *37*(1–3), 69–76. https://doi.org/10.1080/19443994.2012.661256

López-Ramírez, J. A., María Quiroga, J., López-Ramírez, J. A., Oviedo, C., & Quiroga Alonso, J. (2005). Comparative studies of reverse osmosis membranes for wastewater reclamation. *Elsevier*, *191*, 137–147. https://doi.org/10.1016/j.desal.2005.08.013

Lu, S. M. (2018). A global review of enhanced geothermal system (EGS). *Renewable and Sustainable Energy Reviews*, *81*(June 2017), 2902–2921. https://doi.org/10.1016/j.rser.2017.06.097

Missimer, T. M., Choon Ng, K., Thuw, K., & Wakil Shahzad, M. (2016). Geothermal electricity generation and desalination: an integrated process design to conserve latent heat with operational improvements. *Desalination and Water Treatment*, *57*(48–49), 23110–23118. https://doi.org/10.1080/19443994.2016.1144693

Nexus, D. C.-W.-E. et al. (2018). Water for the millions: focus Saudi Arabia. *Water-Energy Nexus*. https://www.sciencedirect.com/science/article/pii/S2588912518300304

Ophir, A. (1982). Desalination plant using low grade geothermal heat. *Desalination*, *40*(1–2), 125–132. https://doi.org/10.1016/S0011-9164(00)88674-9

Ophir, A., & Lokiec, F. (2005). Advanced MED process for most economical sea water desalination. *Desalination*, *182*(1–3), 187–198. https://doi.org/10.1016/j.desal.2005.02.026

Panagopoulos, A. (2021). Water-energy nexus: desalination technologies and renewable energy sources. *Environmental Science and Pollution Research*, *28*(17), 21009–21022. https://doi.org/10.1007/s11356-021-13332-8

Petrovic, A., Delibasic, B., Filipovic, J., Petrovic, A., & Lomovic, M. (2018). Thermoeconomic and environmental optimization of geothermal water desalination plant with ejector refrigeration system. *Energy Conversion and Management*, *178*(October), 65–77. https://doi.org/10.1016/j.enconman.2018.10.035

Pillai, I. R., & Banerjee, R. (2009). Renewable energy in India: Status and potential. *Energy*, *34*(8), 970–980. https://doi.org/10.1016/j.energy.2008.10.016

Prajapati, M., Shah, M., & Soni, B. (2021a). A review of geothermal integrated desalination: A sustainable solution to overcome potential freshwater shortages. *Journal of Cleaner Production*, *326*, 129412. https://doi.org/10.1016/j.jclepro.2021.129412

Prajapati, M., Shah, M., & Soni, B. (2022). A review on geothermal energy resources in India: past and the present. *Environmental Science and Pollution Research*, *29*(45), 67675–67684. https://doi.org/10.1007/s11356-022-22419-9

Prajapati, M., Shah, M., & Soni, B. (2023). Performance evaluation of a novel geothermal energy integrated single effect evaporator desalination with software simulation. *Journal of Cleaner Production*, *407*, 137087. https://doi.org/10.1016/j.jclepro.2023.137087

Prajapati, M., Shah, M., Soni, B., Parikh, S., Sircar, A., Balchandani, S., Thakore, S., & Tala, M. (2021b). Geothermal-solar integrated groundwater desalination system: Current status and future perspective. *Groundwater for Sustainable Development*, *12*(October 2020), 100506. https://doi.org/10.1016/j.gsd.2020.100506

Puppala, H., & Jha, S. K. (2018). Identification of prospective significance levels for potential geothermal fields of India. *Renewable Energy*, *127*, 960–973. https://doi.org/10.1016/j.renene.2018.04.075

Puppala, H., & Jha, S. K. (2021). Extraction schemes to harness geothermal energy from Puga geothermal field, India. *Energy Sources, Part A: Recovery, Utilization and Environmental Effects*, *43*(15), 1912–1932. https://doi.org/10.1080/15567036.2019.1668086

Rybach, L., & Kohl, T. (2018). Geothermal Energy and a Future Earth. *Global Change and Future Earth: The Geoscience Perspective*, 364–376. https://doi.org/10.1017/9781316761489.035

Saeed, M. O., Ershath, M. M., & Al-Tisan, I. A. (2019). Perspective on desalination discharges and coastal environments of the Arabian Peninsula. *Marine Environmental Research, 145*(June 2018), 1–10. https://doi.org/10.1016/j.marenvres.2019.02.005

Shah, M., Sircar, A., Shaikh, N., Patel, K., Thakar, V., Sharma, D., Sarkar, P., & Vaidya, D. (2018a). Groundwater analysis of Dholera geothermal field, Gujarat, India for suitable applications. *Groundwater for Sustainable Development, 7,* 143–156. https://doi.org/10.1016/j.gsd.2018.05.002

Shah, M., Vaidya, D., & Sircar, A. (2018b). Using Monte Carlo simulation to estimate geothermal resource in Dholera geothermal field, Gujarat, India. *Multiscale and Multidisciplinary Modeling, Experiments and Design, 1*(2), 83–95. https://doi.org/10.1007/s41939-018-0008-x

Shelare, S., Kumar, R., Gajbhiye, T., & Kanchan, S. (2023). Role of Geothermal Energy in Sustainable Water Desalination—A Review on Current Status, Parameters, and Challenges. *Energies, 16*(6). https://doi.org/10.3390/en16062901

Singh, H. K., Chandrasekharam, D., Trupti, G., Mohite, P., Singh, B., Varun, C., & Sinha, S. K. (2016). Potential Geothermal Energy Resources of India: A Review. *Current Sustainable/Renewable Energy Reports, 3*(3–4), 80–91. https://doi.org/10.1007/s40518-016-0054-0

Singh, H. K., Thankappan, A., Mohite, P., Sinha, S. K., Chandrasekharam, D., & Chandrasekhar, T. (2018). Geothermal energy potential of Tulsishyam thermal springs of Gujarat, India. *Arabian Journal of Geosciences, 11*(6). https://doi.org/10.1007/s12517-018-3501-y

Sircar, A., Shah, M., Vaidya, D., Sahajpal, S., Chaudhary, A., & Dhale, S. (2017). *Geophysical Techniques for Geothermal Exploration OVERVIEW OF GEOTHERMAL SURFACE EXPLORATION METHODS. May.* https://www.researchgate.net/publication/317178649

Sircar, A., Shah, M., Sahajpal, S., Vaidya, D., Dhale, S., & Chaudhary, A. (2015). Geothermal exploration in Gujarat: case study from Dholera. *Geothermal Energy, 3*(1), 1–25. https://doi.org/10.1186/s40517-015-0041-5

Sircar, A., Solanki, K., Bist, N., & Yadav, K. (2022). Enhanced Geothermal Systems – Promises and Challenges. *International Journal of Renewable Energy Development, 11*(2), 333–346. https://doi.org/10.14710/IJRED.2022.42545

Sircar, A., Yadav, K., & Sahajpal, S. (2007). Overview on Direct Applications of Geothermal Energy. *International Advanced Research Journal in Science, Engineering and Technology ISO, 3297*(9). https://doi.org/10.17148/IARJSET.2016.3925

Stefansson, V. (2000). The Renewability of Geothermal Energy. *Proceedings World Geothermal Congress 2000,* 883–888. https://www.geothermal-energy.org/pdf/IGAstandard/WGC/2000/R0776.PDF

Systems, E. G. (2006). The Future of Geothermal Energy. *Technology, Im*(November), 372. http://www.eere.energy.gov/geothermal/pdfs/structure_outcome.pdf

Torres-Rodriguez, V. (2000). Geothermal Chart of Mexico scale 1: 2000000. *Researchgate.Net.* https://www.researchgate.net/profile/Vicente-Torres-6/publication/352038274_GEOTHERMAL_CHART_OF_MEXICO_SCALE_12000000/links/60b66e6b299bf106f6f1668d/GEOTHERMAL-CHART-OF-MEXICO-SCALE-12-000-000.pdf

Yadav, K., Gudjonsdottir, M., Axelsson, G., et al. (2023). Geothermal-solar integrated Multistage Flash Distillation Cogeneration system: A cleaner and sustainable solution. *Desalination.* https://www.sciencedirect.com/science/article/pii/S0011916423005295

Yadav, K., & Sircar, A. (2021). Geothermal energy provinces in India: A renewable heritage. *International Journal of Geoheritage and Parks, 9*(1), 93–107. https://doi.org/10.1016/j.ijgeop.2020.12.002

Solar Energy Integrated Desalination

4

4.1 INTRODUCTION

Water scarcity has become one of the most significant environmental concerns in recent decades (Abedin et al., 2019). This is a direct outcome of the significant increase in the demand for water on a global scale that has occurred over previous decades (He et al., 2021). The demand for water has been growing at a rate that is larger than the rise in the world population throughout that period, and it is anticipated that this trend will continue due to the increasing demands from the residential, agricultural, and industrial sectors (Manju & Sagar, 2017). The increasing gap between the amount of available water and the amount being used has led to an increase in water scarcity. It has brought about further concerns for the economic and social growth of today's world. As a consequence of this, the availability of freshwater is now considered an essential concern in many regions of the world (Damania et al., 2017). There has been a significant increase in the amount of pressure placed on the quantity and quality of freshwater due to the inefficiency of water supply systems and the contamination and unsuitable utilization of groundwater sources.

The usage of traditional sources of water will not be sufficient to meet the worldwide water demand, either in terms of quality or quantity, given that it is anticipated that water shortages will double by the year 2050 (Vigil, 2023). The desalination process is effective for increasing water supply, particularly when treating non-traditional water sources such as industrial waste, saltwater, brackish water, and other similar water sources. On the other hand, the treatment of non-traditional water sources presents several obstacles regarding feasibility and sustainability because the difficulties are related to expenses, energy intensity, and environmental influence (Goh et al., 2017).

There are 16,000 desalination plants in various stages of development or approval as of 2019. The combined daily freshwater capacity of these plants is 95 million m^3. A total of 13,567 desalination facilities are either operational, expected to be operational, or in the process of being built (Jones et al., 2019). The combined daily freshwater

DOI: 10.1201/9781003530893-4

capacity of these plants is 121 million m^3. This indicates that the desalination operations will experience growth of at least 12% during the next few years. Nearly 90% of facilities use RO membranes for water treatment, making it the most widely used desalination technology today. RO is followed by multi-effect desalination (MED), electrodialysis (ED), and multi-stage flash (MSF) with 6%, 3%, and 2% concentrations, respectively. Nearly 70% of the plants handle brackish water (3000 to 20,000 ppm) or seawater (20,000 to 50,000 ppm) as feedwater. Currently, 21% of the plants handle feedwater with low salinity (TDS < 3000 ppm), whereas 9% handle wastewater. Only 1% of the desalination units in operation today handle concentrated seawater or brine (Caldera & Breyer, 2020).

Almost 99% of the desalination facilities nowadays run on fossil fuels. Should this trend continue, the desalination sector will account for 400 million ton of CO_2 annually by 2025 (Esmaeilion, 2020). Additionally, producing freshwater from desalination creates high-salinity brine (approximately two times above saltwater salinity levels) with residual compounds from the desalination operations. Through a deadly osmotic shock (algae, fish, plankton, seagrass, and others), rejection of the surroundings can damage local ecosystems. Projections show that by 2025, this waste brine will rise to 240 km^3 annually (Azizi et al., 2022). It takes more energy to desalinate brackish or seawater than other water production methods. The energy needed to purify water from rivers and lakes is 0.37 kWh/m^3. The energy needed to cleanse wastewater is between 0.62 and 0.87 kWh/m^3, while the energy required to purify saltwater is between 2.6 and 8.5 kWh/m^3 (Esmaeilion, 2020).

One promising strategy for reducing greenhouse gas emissions is to use renewable energy sources (RES) like solar power to power desalination facilities (Reif & Alhalabi, 2015). Many researchers and experts have been drawn to using solar energy for water desalination due to its accessibility and abundant solar radiation (Coffrin et al., 2008). More and more methods are being developed to harness solar energy and boost freshwater production (Meratizaman & Godarzi, 2019). Desalination mainly employs one of two methods for integrating solar energy: direct systems or indirect systems. Photovoltaics, porous media, nanotechnology, phase transition materials, and localized thermal heating are ways that water can be separated from salts by increasing its evaporation rate (Li et al., 2020).

The chapter discusses the increasing global water scarcity due to higher industrial, agricultural, and residential demand. It highlights the limitations of traditional water sources and the potential of desalination, particularly solar-geothermal hybrid systems, to address water shortages. The chapter reviews desalination technologies' current state, energy requirements, and environmental impacts, emphasizing the need for sustainable and energy-efficient solutions. It explores integrating renewable energy, especially solar power, with desalination processes to reduce greenhouse gas emissions and improve feasibility (Alhaj et al., 2017). The chapter also examines various solar desalination technologies, their efficiencies, and the challenges in scaling production. It concludes by stressing the importance of interdisciplinary research and innovation in renewable energy and desalination to develop integrated systems that provide a reliable water supply while minimizing environmental harm.

Section 4.2 discusses integrating solar energy into desalination systems. It explores various solar desalination technologies, including direct and indirect systems, and their potential for large-scale freshwater production. The section highlights the importance of renewable energy, particularly solar power, in reducing the environmental impact of desalination processes, which traditionally rely on fossil fuels and contribute to CO_2 emissions and brine production. The section underscores the growth of desalination operations and the challenges faced, advocating for interdisciplinary efforts to achieve sustainable water supply solutions. Section 4.3 of the chapter focuses on the direct application of solar energy in desalination processes, mainly through humidification-dehumidification (HDH) systems and solar stills. Solar stills, one of the oldest methods, use solar energy to evaporate saline water, which then condenses to form distilled water. Various enhancements, such as integrating solar stills with other technologies like solar ponds and evacuated tube collectors (ETCs), have improved their efficiency. The section also discusses the factors affecting the performance of solar stills, including the angle of sunlight, water depth, and design configurations. The efficiency of HDH systems is influenced by the temperature of the glass, air, and water, as well as the air velocity within the system. Both methods promise sustainable desalination but require further optimization for large-scale freshwater production. Section 4.4 of the chapter discusses the indirect application of solar energy in desalination processes, focusing on MSF, MED, and reverse osmosis (RO). The section emphasizes the potential of solar energy to power these desalination technologies, reducing reliance on fossil fuels and enhancing sustainability. It advocates for further research and optimization of solar integrated desalination systems to meet global potable water demands.

4.2 SOLAR INTEGRATED DESALINATION SYSTEM

There are many ways in which desalination and renewable energy can be combined (Ghaffour et al., 2015). For integrated designs to succeed, researchers in renewable energy and desalination must work together. Powering the desalination process using energy generated locally from renewable sources is a viable option (Pouyfaucon & García-Rodríguez, 2018). Many forms of energy can be converted into electricity, heat, or mechanical energy. It is theoretically possible for water desalination plants to coexist with any RES, particularly those that use energy for desalination. Suppose power generation is not continuous or stable throughout the day or season, as with solar or wind power systems; linking these facilities to the power grid is an excellent way to offset the energy needed by desalination (Fernandez-Gonzalez et al., 2015). Energy expenditure and production are thus balanced.

A large number of desalination technologies rely on either electrical or thermal input that is harnessed from solar energy (Zuo et al., 2019). As a result of the depletion of fossil fuels and their negative impacts on the environment, much focus has been placed on solar-based systems. Some examples of these systems include photovoltaic (PV) systems and concentrated solar thermal (ST) collectors (El Haj Assad et al., 2021).

Two of the many ways that the sun's thermal energy, which is received as radiation, can be harnessed are concentrated solar power (CSP) (Mohammadi et al., 2019) and PV/concentrator photovoltaic (CPV) solar modules (Elfaqih et al., 2024), which convert the energy that the sun's photons carry into electricity. The efficiency of ST collectors, such as flat plate collectors (FPC), ETCs, and parabolic trough collectors (PTCs), is significantly greater than that of PV systems. The PV system uses solar cells to transform sunlight into usable electricity. Efficiency ranges from 8% to 12% for PV collectors. Concentrating ST collectors use the sun's rays to generate heat, while PV panels use crystalline silicon, thin films, or semiconductor thin films (Goetzberger et al., 2003).

The most often used methods include MSF, multiple-effect distillation (MED), and RO (Rognoni, 2007). In RO, the feed saltwater is subjected to high-pressure compression during the membrane desalination process (Manolakos et al., 2009). An osmotic pressure difference between two solutions occurs when a semipermeable barrier separates a concentrated salt solution from pure water. The osmotic pressure is proportional to the saltiness of the water (Mayere, 2011). To maintain a chemical equilibrium between the two solutions, it is common for pure water to pass through the membrane and move from the solution with a lower concentration to the solution with a higher concentration (Heidary et al., 2018). When a pressure that is higher than the osmotic pressure is applied to the concentrated solution, the membrane can separate the pure water from the solution that contains salt. The energy efficiency of this desalination technology is unparalleled. At 50% recovery from 35 g/L of seawater, it uses 2–8 kWh/m^3 of power (the thermodynamic limit is approximately 1 kWh/m^3) and costs less than \$1/m^3 (Shrivastava et al., 2015).

The method of MSF distillation involves boiling saline water followed by condensing it in a sequence of vessels, referred to as stages (Heidary et al., 2018). This technique is employed to acquire potable water. Each vessel contains tubes through which the feed water circulates and is heated by the condensation energy of the water vapor generated. The materials undergo thermal treatment in a brine heater and are then added to the main vessel, where the pressure is reduced. The reduction in pressure leads to the fast evaporation of a fraction of the pure water in the saline solution, referred to as the flashing phenomenon (Belessiotis et al., 2016). The vapor is cooled via condensation onto the feed water tubes and then collected. The residual portion of the solution experiences a following phase at a much-reduced pressure, resulting in its quick vaporization once again, and this cycle is repeated (Reif & Alhalabi, 2015).

ST collectors can harness sun energy for heat production in thermal desalination processes (Belessiotis et al., 2016). Devices driven by solar energy have seen substantial advancement and enhancement in recent decades. The primary limitation of most ST desalination systems, which rely on an HDH (Orfi et al., 2004), solar still design (Panchal & Shah, 2012), and membrane distillation processes, is their restricted capacity for water production rates. Particularly in the most remote rural areas, they are quite beneficial.

Both direct and indirect methods exist for harnessing solar energy, but they differ in the specifics (Reif & Alhalabi, 2015). Solar stills and other systems that desalinate water directly from the sun capture ST energy (Yadav & Sudhakar, 2015). In contrast, ST or electrical energy can power indirect solar desalination systems. Although there have been several optimizations and state-of-the-art research on solar still technologies, they are still not widely used to produce substantial amounts of freshwater. More advanced solar stills for desalination are also the subject of active study. The term "indirect desalination system" refers to various methods, including humidification and dehumidification, ED, MSF, MED, and RO. While RO may work for large-scale desalination projects, it is unsuitable for smaller-scale operations (Vishwanath Kumar et al., 2015).

4.3 DIRECT APPLICATION

The utilization of solar energy directly allows for the simultaneous collection and desalination of salt water. Solar stills and HDH systems are frequently used since they rely on straightforward mechanisms and avoid complicated phenomena. Researchers are focusing on this area and looking at ways to improve it as the satisfaction level did not match the productivity and efficiency (Pourkiaei et al., 2021). The results of these experiments show that the efficiency and productivity of direct desalination devices can be affected by several factors, including the angle of incidence of the sun's rays, the depth of the water in the basin, the transmissivity of the cover plate, the absorption of the basin line, the device's size, shape, and the wind speed, orientation, and flow rate, and the ambient temperature and humidity (Chauhan et al., 2021).

4.3.1 Solar Stills

Solar stills are among the oldest and simplest ways to use solar energy to remove salt from water (H. Panchal et al., 2019). A solar still is a machine that directly evaporates salty water using solar energy. The collected pure water is then stored (Sivaku & Jidhesh, 2016) (Figure 4.1).

A solar still comprises a basin that has been blackened on the inside and filled with brackish or salty water up to a certain depth. This helps the passage of solar energy and condensation. Then, an angled piece of glass is put over the sink. Solar radiation that gets into the container warms up the blackened liner, which warms up the water and causes it to evaporate (Arunkumar et al., 2012). The different temperatures and partial pressure allow the water vapor to condense along the slanted glass cover. It is then collected by a fitting feature at the basin's bottom. Additionally, the condensate obtained will be of a high grade and have a daily measurement of between $2L/m^2$ and $3L/m^2$ (Kapnichor, 2017).

The condensate yield has been enhanced by integrating solar stills with a solar pond, ETC, flat plate collector, wind turbine, concentrating dish type collector, reflectors, and air conditioning unit (Ayoub et al., 2015). Furthermore, it has been shown that increasing the quantity of phase change material (PCM) and extending the utilization of the latent heat of condensation by using many effects can improve condensate yield. A reasonable improvement in yield was observed after coating the basin liner with photocatalysts such as MnO_2, PbO_2, and CuO and adding heat-generating media such as iodine and bromine containers (Alghassab, 2024). The condensing cover material, water depth, angle, and the coating on the absorber significantly affect the distillate yield. For minimum water depth, it was discovered that a glass cover acting as a condensation surface at an angle equal to the location's latitude yielded the best results. Solar stills with a single slope or a pyramid shape produce the most distillate daily on average and at the highest possible rate, with water production costs of 0.0135 & 0.031 USD/l, respectively (G. Tiwari et al., 1994).

Several aspects still continue to influence the efficiency and rate of evaporation in solar energy. The intensity of the sun's direct beams is a crucial determinant in optimizing the efficiency of solar stills. Under maximum solar radiation intensity, the distillate production rate is most significant during the early afternoon (Feria-Díaz et al., 2021). The pace at which evaporation occurs is directly related to the exposed area, also known as the active area, of the still. The volume of distillate generated by a solar still is proportional to the water basin's dimension. The depth of the water bath is another consideration that might reduce the efficiency of the still (Ayoub & Malaeb, 2014). Because the solar still's distillate product decreases as the water bath depth increases, increasing the still's daily productivity requires decreasing the water basin depth. A solar still's output can be optimized by adjusting its tilt and latitude angle. Predicting the best angles for increased output, however, is location-dependent (Khan et al., 2018).

The primary obstacle solar energy still faces is the limited efficiency per unit of installation area. The performance of the solar still can be improved by optimizing and adjusting many factors, such as the design configuration, amount of light transmitted, thickness of the glass, angle of the cover, depth of the brine, shape of the still, and control of heat transmission within the solar still (Haddad et al., 2017). Sensible- and latent-heat storage materials can be combined with solar-still desalination to use the energy during periods of low sunlight or cloudy weather. Furthermore, nanofluids can be utilized to augment the thermal conductivity of water, resulting in an elevation in the absorption rate of solar radiation, hence enhancing the productivity of solar stills (Manchanda & Kumar, 2015). Using nanofluids improved the performance of a solar still with an external condenser by around 116%. The thickness of the glass cover is a crucial factor that significantly influences the heat transfer rate by serving as the primary medium for incoming radiation. A glass cover plate with a thickness of 3 mm achieves the highest production rate of 15.5%, while increasing the thickness decreases the overall production (Hilarydoss, 2023).

Kannakumar et al. (2023) and Velliangiri et al. (2013) conducted a comprehensive evaluation of the several factors that influence the efficiency of solar stills. The main variables included the choice of preheating feed water, wick materials, PCM, heat storage, and their combination with FPC and ETCs to achieve the most efficient distillation process. Al-Harahsheh et al. (2018) conducted an empirical investigation on solar stills, specifically focusing on utilizing PCM to store latent heat. Consistent with a continuous energy transfer process, the PCM stores energy from solar radiation during the day and releases it at night. The results showed that after sunset, at a rate of 4.3 L/m^2, 40% of the water was desalinated (Al-Harahsheh et al., 2022). An experimental study on solar stills coupled with thermal heat storage was carried out by Kabeel and El-Said (2013). A mechanism for thermal heat storage was built using paraffin wax and graphite nanoparticles. Results showed a maximum thermal efficiency of 65.13% and a water production rate of 8.5 L/m^2/day.

El-Agouz et al. (2015) presented a theoretical model for analyzing the stepped solar still. They did this by detailing the energy balance for four areas: the glass cover, the saline water, the basin, and the cooling water film. For film water cooling analysis, it is necessary to be aware of the temperatures at which the water enters and exits the film, as the authors have pointed out. Their prior experimental work is in excellent agreement with the theoretical model they constructed. Based on their numerical investigation, Mousa found that water film cooling improved the solar still's efficiency by 20% compared to film cooling settings alone.

Gnanadason et al. (2015) conducted a novel study on solar stills, utilizing a floater absorber to produce water. This investigation involved the application of a soot solution to a cotton-based absorber to replicate the capillary evaporation process. An experimental investigation was conducted to assess the effectiveness of the recommended absorber. The results indicated that a photothermal efficiency of 80% can be achieved. Tests for phosphate, nitrate, and sulfate levels confirmed that the extracted water was safe for drinking.

Rajaseenivasan and Srithar (2016) enhanced the daily efficiency of solar stills by incorporating circular and square fins in the basin. The findings validated that employing square fins in the solar still basin has substantially enhanced productivity compared to circular fins. The square fins in a solar still led to an enhanced total productivity of 4.6 kg/m^2, surpassing the traditional solar still.

Saxena et al. (2013) enhanced the efficiency of a solar still by incorporating porous fins in the basin. The fins serve as an expanded surface to increase efficiency and as a means of storing thermal energy. The outcomes indicated that the production increased to 7.5 kg/m^2, surpassing the traditional solar still under summer climate conditions.

An empirical investigation was carried out by Sansom and Tonnellier (2018) on integrating solar stills with parabolic dish collectors. A parametric study evaluated the impact of critical parameters such as reflectivity, efficiency, collector optical, and aperture area. Analysis revealed that decreasing the reflectance of the parabolic dish to 0.4 results in an improvement in optical efficiency to 0.8. Hence, there is a rise in the generation of desalinated water.

Kalbasi et al. (2017) developed a theoretical framework for the solar still, considering both single and double effects. The experimental analysis confirmed the theoretical results. The amount of distilled water produced depends on the temperature of the water in the basin and the difference in temperature between the surface at which the water condenses and the basin water. The study determined that the solar still's production increased by approximately 94% compared to the conventional still. This improvement was achieved by separating the condensing surface from the surface exposed to sun radiation.

In addition to deriving a transient thermal model, Naroei et al. (2018) created a stepped solar still connected to a photovoltaic thermal water collector (PVT). For the elemental temperatures, the numerical findings showed an average error percentage between 5.76 and 6.666%. Their research states that more than twice as much energy was saved, and freshwater production increased by 20% because of the PVT collector.

A. Tiwari and Tiwari (2006) studied passive and active solar distillation technologies to provide rural communities access to clean water and suggested using passive solar stills with a twice slope. Compared to stationary solar stills, those using a computerized single-axis sun tracking system that moves from east to west saw a 22% increase in water production.

Suneesh et al. (2017) conducted a detailed analysis to determine the thermal efficiency of a V-type solar still that employed a charcoal absorber and collected distilled water. Detailed analysis is carried out on both the external heat transfer and the internal heat transfer mechanisms. The system's effectiveness remains at 24.47% in the absence of charcoal, 30.05% when charcoal is present, 11.92% when a boosting mirror is utilized, and 14.11% when both a boosting mirror and charcoal are jointly used.

Elango et al. (2015) conducted an experimental investigation on the performance of a double slope solar still with a 30° angle on a basin-type design. They examined the effects of various wick materials and the minimal water required. The effectiveness of black cotton fabric combined with a still has been determined. The highest possible output can be attained using a rectangular fin coated with cotton cloth in the longitudinal direction.

The thermal efficiency of a solar still combined with an ETC was investigated by Mamouri et al. (2014). The experimental configuration comprises a basin, five ETCs, and thermosyphon heat pipes. A small, easily transportable solar still with a single solar effect is linked to an ETC in order to increase the rate at which water evaporates onto the glass condensation surface. Different coverings made of glass, steel, and aluminum were used to evaluate the effectiveness of the solar still. The glass cover showed the highest success rate and effectiveness in producing distilled water when a water depth of 2 cm was employed, which aligned with the ideal dimensions of the condenser section of the heat pipes. The peak yield and efficiency were 1.02 L/m^2/h and 22.9%, respectively. The Cost Per Liter (CPL) factor over the unit area is 0.0092 USD/L.

Eze et al. (2012) compared solar stills of different shapes. Additionally, in the coastal climates of Nigeria, Eze and Ojike compared two passive solar stills with a single basin. Kalaivani et al. (2012) built a pyramid solar still out of stainless steel, measuring 0.75 m × 0.75 m × 0.20 m. The effectiveness of solar stills shaped like triangles was investigated by Kabeel et al. (2017). Moreover, Sampathkumar et al. (2010) introduced a new idea for improving the efficiency of solar stills fashioned like pyramids. They integrated the biomass-powered heater with the pyramid solar still. Kabeel et al. (2017) experimented in Tanta City, Egypt, to measure the angles of tilt of the condensation surface of a square pyramid solar still. Kalogirou (2005) also reviewed several different types of solar collectors. Also included were comprehensive technical specifications for various collectors in thermal, water heating, desalination, cooling, and space heating applications.

Figure 4.1 illustrates the components and functioning of a solar still system for efficient energy use.

In most published work, many review articles concentrate on developments in solar still productivity. These publications cover changing operating circumstances, design

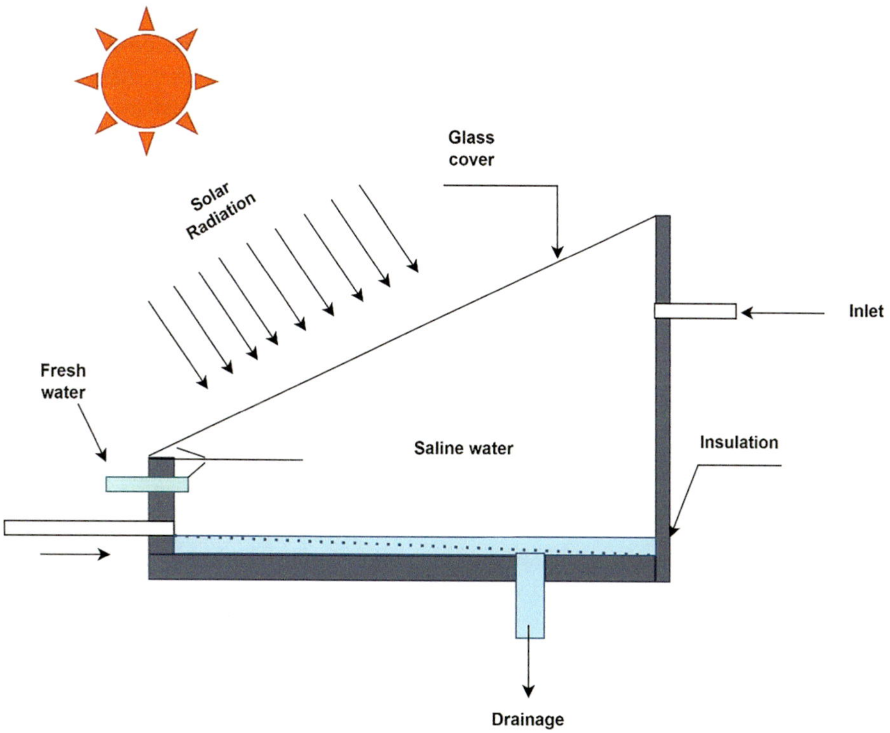

FIGURE 4.1 Solar still.

configurations, and solar stills that include latent heat storage and may be utilized when there is no sunshine (Alam et al., 2023; Eltawil et al., 2009). Solar designs that are still available are not yet suitable for extensive systems. Even though solar stills require little maintenance and are inexpensive, they have low efficiency and significant heat loss, even for smaller systems, and their production is restricted.

4.3.2 Solar-powered Humidification-dehumidification Desalination

When the sun heats the water, a natural hydrological process is analogous to humidification and dehumidification. After that, clouds are formed as a result of the water in the ocean evaporating due to a process known as natural convection. Rainfall results from the cloud condensing and releasing water into the atmosphere (Capocelli et al., 2023).

Humidification and dehumidification are additional methods by which salty water can be converted into freshwater through solar power. This method simulates the natural rain cycle for a shorter time than the natural cycle (Fthenakis et al., 2022). The humidifier and the dehumidifier are situated in two different locations, and the air is the medium via which water vapor is transported. Solar irradiation causes the dry air in the atmosphere, which may also be moist air, to heat up and become saturated as it travels through the basin space between the raw water and the glass cover (Jackson et al., 2022). Subsequently, after traveling along the basin's length, this saturated air condenses at the glass cover at the appropriate pressure and temperature. Typically, condensation on the surface of a glass film causes it to become moist, leading to a further reduction in the rate of condensation. The liquid condenses and flows down to the trough due to the force of gravity. The temperature of the air, glass, and water mainly influences the efficiency of the HDH system (Lawal et al., 2018).

Additionally, there is a negative correlation between the internal air velocity of the system and the level of productivity achieved. Water vapor, humidity, or moisture can be introduced into the air by humidification, which does not change the air's dry bulb temperature (DBT) (Nafey et al., 2004). In contrast, dehumidification takes place on substantially colder surfaces. This method has been considered reliable and efficient in acquiring significant quantities of potable water through direct desalination (Parekh et al., 2004).

The fundamental concept underlying the solar HDH process is that the air's ability to hold moisture expands as the temperature rises. When circulated in either a natural or forced mode, solar collector-heated hot air absorbs a certain quantity of vapor when it comes into touch with salty water sprayed in the evaporator. The condenser can recover this vapor where the saltwater feed water is heated (A. Tiwari & Kumar, 2023).

An empirical investigation was carried out on a pilot plant for solar desalination at INRST Tunisia. The innovative approach employs an air multiple-effect humidification and dehumidification technology to produce an average of 355 kg/day of freshwater in August. The highest yield attained is 516 L/day. However, the expense of this procedure is somewhat excessive, amounting to 28.65/m^3 (Houcine et al., 1999).

Gabrielli and Mazzotti (2019) developed an HDH solar desalination system in 2019, integrated with a photovoltaic thermal water collector (PVT) system. Additionally, a techno-economic study was conducted. Consequently, a humidification column was constructed alongside a heat exchanger, and a suitable mathematical relationship was introduced. The findings demonstrated that the mean water production can be achieved at a rate of 0.21 kg/h m^2.

Hosseini and Sarhaddi (2017) proposed a desalination technique known as open water closed air (OWCA) HDH, where salt water is heated in a dehumidifier before being introduced into a PVT panel and humidifier. This study aimed to assess the influence of the quantity of PVT panels on the system's performance. Enhancing the quantity of PVT collectors maximizes desalination efficiency and power generation. However, it increased the energy consumption required for pumping and decreased total efficiency. An energy efficiency of 65% was attained under optimal conditions of a saline water mass flow rate of 0.025 kg/s and a humid air mass flow rate of 0.03 kg/s (Figure 4.2).

Figure 4.2 Schematic of a solar energy integrated HDH system, showcasing the various stages involved in the treatment of seawater.

Giwa et al. (2016) researched integrating a PVT panel with OWCA-type high-frequency desalination. While the air passage is responsible for cooling the PVT panel, the hot air is then routed to the humidifier and dehumidifier before being directed back to the PVT panel, thereby completing a full loop. According to the study's findings, the combination of PV-RO desalination and PVT-HDH desalination is a more

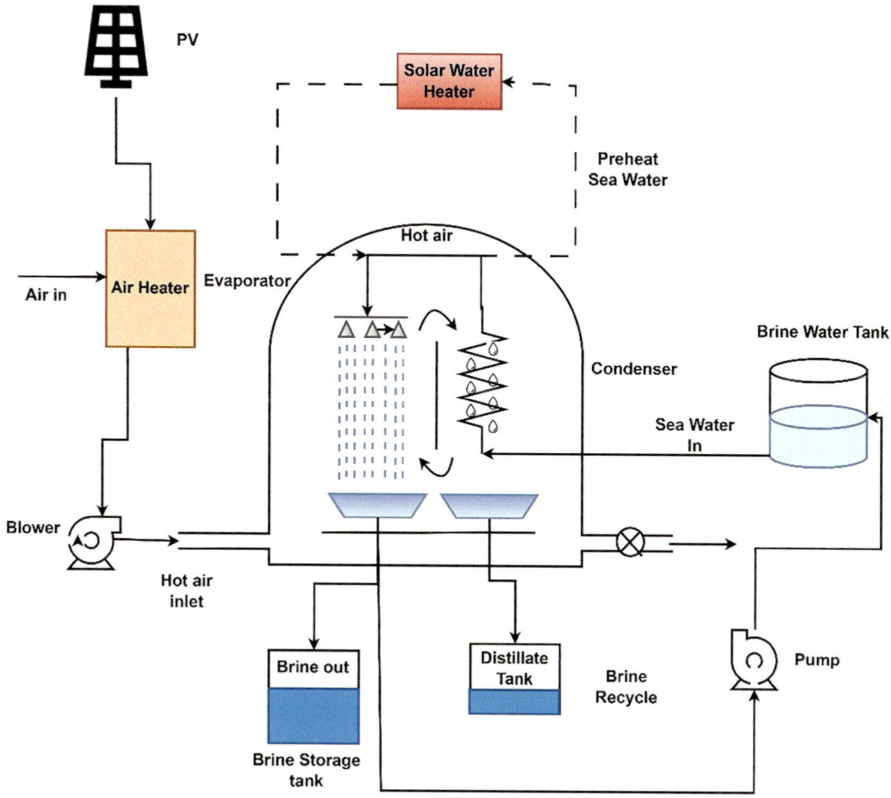

FIGURE 4.2 Solar energy integrated HDH system.

environmentally friendly and practical approach for producing freshwater on its own. Compared to the PV-RO desalination process, the average daily desalination capacity of the PVT panel was 2.28 L/m^2, which resulted in an impressive 83.6% reduction in the environmental effect.

Yuan et al. (2011) integrated theoretical research with practical testing to provide a thorough study on a PV panel integrated with a water-heated HDH desalination system that is brackish water-cooled. Integrating PVT technology with HDH desalination was the first step in the investigation. A PV electric heater was set up in the evaporation system. Afterward, brackish water was used to cool and dehumidify the PV panel. The 3.5 kW power rating of the PV module was sufficient to operate auxiliary equipment such as a pump, fan, and electric heater. They tested the system at various process temperatures utilizing natural and forced convection modes. With a daily production of 0.873 L/m^2, the forced convection mode produced the most freshwater. The cost of freshwater was \$2.3/m^3 when using natural convection and \$21.8/m^3 when using forced convection. At a brackish water evaporation temperature of 64.3°C, the optimal operating condition for supplying freshwater to a family of four adults (2 L per adult per day) through forced convection is 9.16 m^2 of PV panel area.

Gabrielli and Mazzotti (2019) described a PVT-HDH desalination process that produces power and freshwater simultaneously. A comparative analysis was performed to evaluate the suggested system in relation to a conventional ST collector-driven HDH unit. Based on the research findings, it can be concluded that PVT panels offer the most significant economic benefits in regions with high energy expenses. In contrast, thermal collectors present a financially viable alternative in regions with low power costs. This study demonstrates that PVT-HDH desalination is a very effective and economical method for independent desalination in areas characterized by abundant solar radiation and costly energy. This approach proves to be especially effective when taller humidifiers are employed.

4.4 INDIRECT APPLICATION

4.4.1 Multi-effect Desalination

MED is a thermal desalination process that involves raising the temperature of seawater while it is under pressure (Mabrouk & Fath, 2015). In this process, the saltwater is forced to move through a succession of chambers, also known as stages. Solar energy can obtain the heat necessary to cause evaporation in the first stage. The vapors generated during this process are then employed to raise the feed water temperature in the following chamber or stage. Some of the feed water may undergo evaporation, leading to a somewhat more saline solution than the original saltwater (L. Kumar et al., 2023). However, this approach involves harnessing the water vapor from each chamber to warm the water in the subsequent area with a lower pressure than the preceding chamber. Consequently, this ensures that the water is heated to the appropriate temperature. While this repeating procedure is implemented in every succeeding chamber to enhance the system's overall efficiency, the primary mechanism utilizes the heat produced by condensation to provide heat to the next batch of seawater (Mohammed & Askalany, 2019). Consequently, the process leads to the isolation of condensed water vapor, namely distilled water, and the generation of further water vapor as the stages proceed (Ullah & Rasul, 2019).

Figure 4.3 Schematic of solar integrated multiple-effect desalination (MED) unit. The system includes solar collectors, which help to heat the thermic fluid circulating through the pipes. Now, this hot thermic fluid is utilized as a heating source, introduced in the first effect of the MED system. After utilizing the heat, it will be further recirculated in the solar collector system.

A solar-powered MED plant's overall design is similar to that of a conventional MED plant Kumar (2017). This system has two main components: the solar field component and the MED unit component (Ullah & Rasul, 2019). An intermediary storage medium can link both blocks, while a heat exchanger transfers ST energy to the first

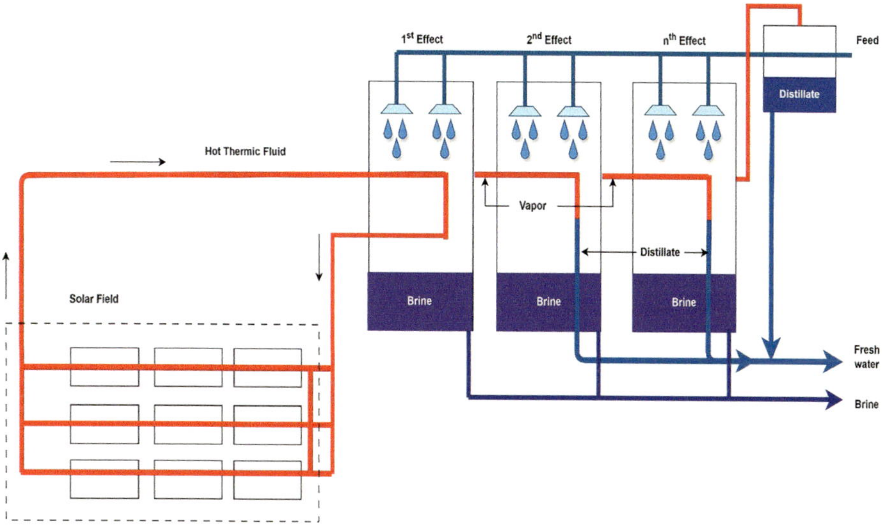

FIGURE 4.3 Solar integrated MED unit.

stage of the MED chamber. Figure 4.3 illustrates a simplified block diagram of a solar-powered MED system. Typically used solar collectors in solar-driven MED are FPCs, linear Fresnel collectors (LFCs), and PTCs (Alhaj et al., 2018).

In addition, several specific research initiatives have investigated the possibility of employing compound parabolic concentrators (CPCs) and solar power towers. The solar field can harvest ST energy by employing a heat transfer fluid (HTF) to capture the energy. This energy is then transported to the MED chamber utilizing a heat exchanger, which is a steam generator. To compensate for any decrease in the intensity of the sun, thermal energy storage (TES) can be utilized (Byrne et al., 2015; Meratizaman & Godarzi, 2019).

Iaquaniello et al. (2014) comprehensively evaluated a MED-RO system that utilizes solar energy from a PTC to generate electricity and create desalinated water. The investigation included both technical and economic aspects. The study utilized a resilient computer model for its simulations. Their hybrid setup led to a water expense of under \$1.23/m^3. A study revealed that extending the plant's lifespan from 20 to 30 years results in an 8% decrease in water expenses. Mabrouk and Fath (2015) introduced the design and simulation of a MED system that a PTC powers. The concept proposed the utilization of brine recirculation in addition to an air-cooled condenser. The study found that using an air-cooled condenser reduces the initial capital outlay needed for groundwater treatment. Sharaf et al. (2011) conducted a thermo-economic assessment of two variations of a solar-powered MED plant, one with a thermal vapor compressor (TVC) and one without. Both configurations were evaluated for their ability to produce desalinated water and generate electricity. The study concluded that producing desalted water alone is more advantageous than cogeneration concerning water expense and solar field area.

4.4.2 Multi-stage Flash

MSF distillation is a process that involves boiling saline water, which utilizes solar energy and then condensing it in a series of vessels, known as stages, ranging from 4 to 40 (Heidary et al., 2018). This procedure leads to the production of water that is free of impurities. Each vessel contains tubes that let the feed water pass through them. The condensation energy of the produced water vapor heats the feed water. In the subsequent step, the feed water is heated in a brine heater and put into the first container, where the pressure is lowered (Bodalal et al., 2010). This process, known as the flashing phenomenon, is caused by a reduction in pressure, which causes a portion of the clean water in the saline solution to evaporate rapidly (Reif & Alhalabi, 2015). Afterward, the vapor is collected once condensed on the feed water tubes. The remaining portion of the solution proceeds to a subsequent phase at an even more reduced pressure, where it undergoes another rapid vaporization process, and this cycle repeats (Morin, 1993). During the 1980s and 1990s, MSF emerged as the most dependable desalination method before advancements in RO membranes (Heidary et al., 2018). The electricity consumption required to operate the Hoover pumps is substantial, ranging from 3.5 to 5 kWh/m^3. Additionally, a significant thermal energy consumption can exceed 40 kWh/m^3. Nevertheless, the price of water remains below \$1.5/m^3 (Nafey et al., 2006).

Figure 4.4 Schematic of solar integrated MSF unit. The system includes solar collectors, which help to heat the thermic fluid circulating through the TES system. Now, this hot thermic fluid is utilized as a heating source for heating secondary fluid, which is introduced in the first effect of the MSF system, followed by the heater.

4.4.3 Reverse Osmosis

RO is a technique that uses pressure to separate freshwater from seawater via a semi-permeable membrane using the solution-diffusion mechanism (Garg et al., 2018). The standard operating pressure for seawater reverse osmosis (SWRO) typically ranges from 55 to 70 bars, but for brackish water reverse osmosis (BWRO), it is lower, ranging from 15 to 30 bars (Al-Bahri et al., 2001). Solar energy utilization to power RO heavily depends on a thorough understanding of RO plant energy consumption rates. Various factors, including the feed-intake facility, pre- and posttreatment, and brine disposal, determine a RO plant's specific energy consumption (SEC) (Manolakos et al., 2009). However, the membrane desalination section accounts for 60–80% of the SEC. This involves the utilization of high-pressure pumps, energy recovery devices, and membrane trains (Figure 4.4) (Leblanc, 2009).

Operating characteristics such as feed salinity and recovery rate permeate quality, and feed temperature influence the need for pressure, affecting energy usage (Anand et al., 2021). Although there have been some enhancements, a substantial proportion of the costs associated with RO still result from the electrical energy needed to pressurize the feed. Electricity or mechanical pumps can supply the applied pressure. Therefore, RO plants can be powered by PV or ST technologies; however, the latter requires either a

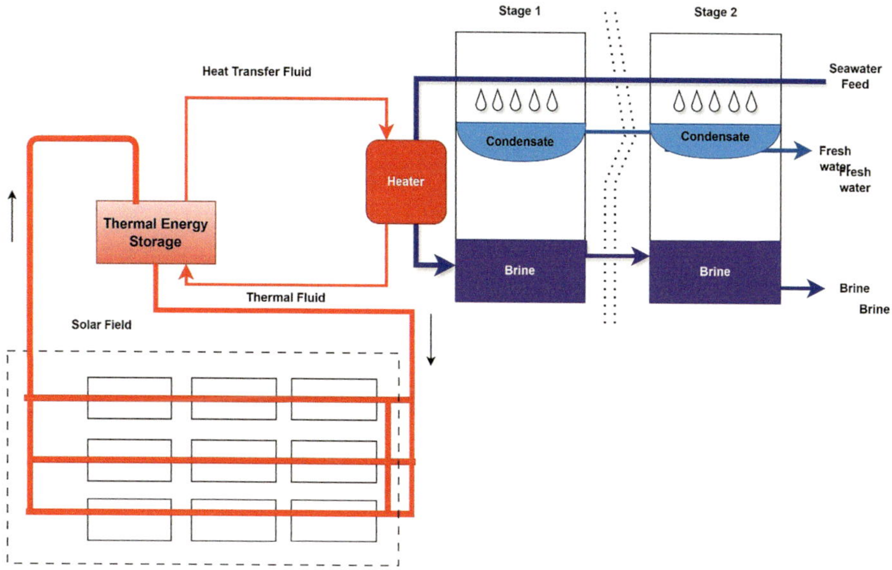

FIGURE 4.4 Solar integrated MSF unit.

power conversion unit or a thermal energy-operated water pump to pressurize the feed (Sharon & Reddy, 2015).

The water production capacities of Al Khafji range from 16 m^3/day to 60,000 m^3/day. Al Khafji is the first desalination plant of its kind in the world to use photovoltaic reverse osmosis (PV-RO) technology. Despite being scheduled to commence in 2017, the plant is now experiencing delays and is still in the process of being built. PV-RO, a technique that combines PV power and RO, has become a well-established and commercially viable option (Schallenberg-Rodríguez et al., 2020).

Raval and Maiti (2015) introduced an innovative method for increasing the efficiency of PV and RO systems. The proposed approach involves harnessing thermal energy from the PV panel and using it to warm the running feed water. This procedure offers the dual advantage of decreasing the temperature of the PV module and increasing the temperature of the RO feed. An investigation revealed that the influence of temperature on PV modules and RO manufacturing can lead to a reduction in energy usage of up to 28%. A 3% increase in membrane flow is seen with each level elevation in feed water temperature (Raval & Maiti, 2016).

Conversely, the electrical efficiency of PV systems decreases as temperature increases, making higher temperatures less favorable. The efficiency of crystalline silicon PV cells decreases by 0.2–0.5% for every degree increase in temperature. The upgraded system resulted in a 20% increase in water output.

Kelly and Gibson (2011) employed concentrated mirrors positioned alongside the PV cell to direct solar radiation onto the cell, resulting in increased electricity generation. The high concentration of radiation-generated heat, which was consistently dissipated by circulating water at the base of the PV plant. The effluent from the concentrated

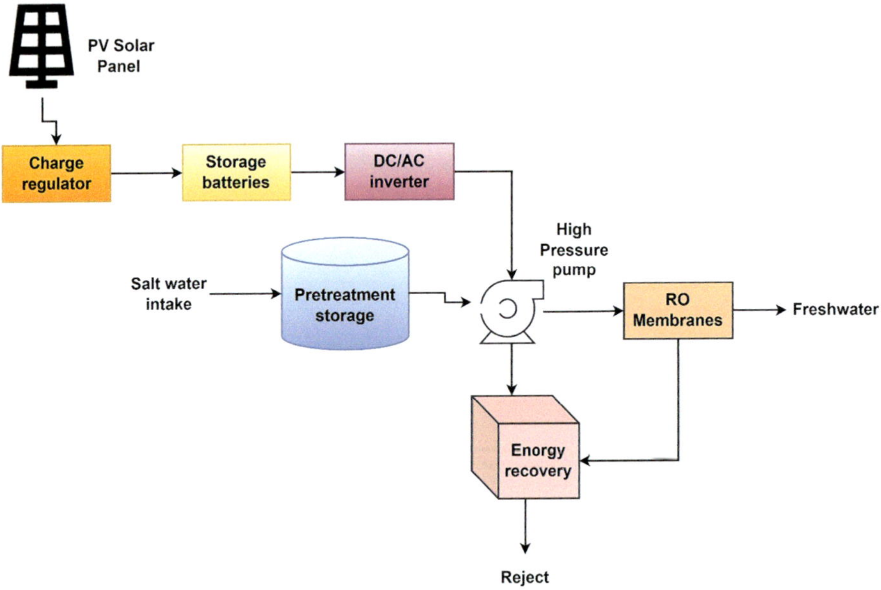

FIGURE 4.5 Solar integrated RO unit.

PV panel was directed to the entrance of the RO process with precise temperature regulation. When equipped with concentrating mirrors, the PVT-RO facility yielded a freshwater output that was 57% greater than that of PV-driven RO desalination.

Figure 4.5 Schematic of solar integrated RO unit, which includes different parts that follow the following stages: PV solar panel, charge regulator, storage batteries, DC/AC inverter, pretreatment unit, high-pressure pump, RO membranes unit.

A performance study of a small PVT-powered RO plant was conducted by Kroiß et al. (2014) Model results showed that when comparing the integrated system to a standalone PV RO plant, the collector area for the latter might be as much as 30% less. In addition, the PVT-RO plant produced more freshwater than 40% of the time. When comparing PVT-RO to PV-RO desalination, the former had a 2% lower specific energy usage.

4.5 CONCLUSION

Integrating solar energy into desalination systems offers a promising solution to address global water scarcity. The chapter provides the advancements in solar-operated desalination technologies, such as HDH and solar stills systems, which have shown significant potential in rising freshwater production. RES, particularly solar energy, contribute to the sustainability of desalination technology and mitigate the environmental impact associated with conventional fossil fuel-based processes. The chapter concludes that adopting solar integrated desalination technologies is crucial for ensuring a reliable and eco-friendly water supply, especially in areas with abundant solar radiation. Future research and development efforts are encouraged to optimize these technologies further and expand their application to meet the increasing global demand for drinking water.

REFERENCES

Abedin, M. A., Collins, A. E., Habiba, U., & Shaw, R. (2019). Climate Change, Water Scarcity, and Health Adaptation in Southwestern Coastal Bangladesh. *International Journal of Disaster Risk Science*, *10*(1), 28–42. https://doi.org/10.1007/s13753-018-0211-8

Alam, J., Daoud, O. A., Shukla, A. K., Ali, F. A. A., & Alhoshan, M. (2023). Simulation of a Solar-Powered Reverse Osmosis System Integrated with Vacuum Membrane Distillation for Desalination Brine Treatment. *Arabian Journal for Science and Engineering*, *48*(12), 16343–16357. https://doi.org/10.1007/s13369-023-08212-x

Al-Bahri, Z. K., Hanbury, W. T., & Hodgkiess, T. (2001). Optimum feed temperatures for seawater reverse osmosis plant operation in an MSF/SWRO hybrid plant. *Desalination*, *138*(1–3), 335–339. https://doi.org/10.1016/S0011-9164(01)00282-X

Alghassab, M. A. (2024). A review of hybrid solar desalination systems: structure and performance. *Water Science and Technology*, *89*(5), 1357–1381. https://doi.org/10.2166/wst.2024.042

Alhaj, M., Hassan, A., Darwish, M., & Al-Ghamdi, S. G. (2017). A techno-economic review of solar-driven multi-effect distillation. *Desalination and Water Treatment*, *90*(January), 86–98. https://doi.org/10.5004/dwt.2017.21297

Alhaj, M., Mabrouk, A., & Al-Ghamdi, S. G. (2018). Energy efficient multi-effect distillation powered by a solar linear Fresnel collector. *Energy Conversion and Management*, *171*(March), 576–586. https://doi.org/10.1016/j.enconman.2018.05.082

Al-Harahsheh, M., Abu-Arabi, M. et al. (2018). Solar desalination using solar still enhanced by external solar collector and PCM. *Applied Thermal Engineering*. https://www.sciencedirect.com/science/article/pii/S1359431117310001

Al-Harahsheh, M., Abu-Arabi, M. et al. (2022). Self-powered solar desalination using solar still enhanced by external solar collector and phase change material. *Applied Thermal Engineering*. https://www.sciencedirect.com/science/article/pii/S1359431122000825

Anand, B., Shankar, R., Murugavelh, S., Rivera, W., Midhun Prasad, K., & Nagarajan, R. (2021). A review on solar photovoltaic thermal integrated desalination technologies. *Renewable and Sustainable Energy Reviews*, *141*(August 2020), 110787. https://doi.org/10.1016/j.rser.2021.110787

Arunkumar, T., Vinothkumar, K., Ahsan, A., Jayaprakash, R., & Kumar, S. (2012). Experimental Study on Various Solar Still Designs. *ISRN Renewable Energy*, *2012*, 1–10. https://doi.org/10.5402/2012/569381

Ayoub, G. M., Al-Hindi, M., & Malaeb, L. (2015). A solar still desalination system with enhanced productivity. *Desalination and Water Treatment*, *53*(12), 3179–3186. https://doi.org/10.1080/19443994.2014.933040

Ayoub, G. M., & Malaeb, L. (2014). Economic feasibility of a solar still desalination system with enhanced productivity. *Desalination*, *335*(1), 27–32. https://doi.org/10.1016/j.desal.2013.12.010

Azizi, N., Esmaeilion, F., Moosavian, S. F., Yaghoubirad, M., Ahmadi, A., Aliehyaei, M., & Soltani, M. (2022). Critical review of multigeneration system powered by geothermal energy resource from the energy, exergy, and economic point of views. *Energy Science and Engineering*, *10*(12), 4859–4889. https://doi.org/10.1002/ese3.1296

Belessiotis, V., Kalogirou, S., & Delyannis, E. (2016). Thermal solar desalination: Methods and systems. In *Thermal Solar Desalination: Methods and Systems* (pp. 1–368). https://doi.org/10.1016/C2015-0-05735-5

Bodalal, A. S., Mounem, S. A. A., & Salama, H. S. (2010). Dynamic Modeling and Simulation of MSF Desalination Plants. *Jjmie*, *4*(3), 394–403.

Byrne, P., Fournaison, L., Delahaye, A., Ait Oumeziane, Y., Serres, L., Loulergue, P., Szymczyk, A., Mugnier, D., Malaval, J. L., Bourdais, R., Gueguen, H., Sow, O., Orfi, J., & Mare, T. (2015). A review on the coupling of cooling, desalination and solar photovoltaic systems. *Renewable and Sustainable Energy Reviews*, *47*, 703–717. https://doi.org/10.1016/j.rser.2015.03.083

Caldera, U., & Breyer, C. (2020). Strengthening the global water supply through a decarbonised global desalination sector and improved irrigation systems. *Energy*. https://www.sciencedirect.com/science/article/pii/S0360544220306149

Capocelli, M., Balsamo, M., Mazzeo, L., & Luberti, M. (2023). Thermodynamic analysis of the humidification-dehumidification-adsorption (HDHA) desalination process. *Desalination*, *554*. https://doi.org/10.1016/j.desal.2023.116510

Chauhan, V. K., Shukla, S. K., Tirkey, J. V., & Singh Rathore, P. K. (2021). A comprehensive review of direct solar desalination techniques and its advancements. In *Journal of Cleaner Production* (Vol. *284*). Elsevier Ltd. https://doi.org/10.1016/j.jclepro.2020.124719

Coffrin, S., Frasch, E., & Santorella, M. (2008). *Solar powered water distillation device* (pp. 1–39).

Damania, R., Desbureaux, S., Hyland, M., Islam, A., Moore, S., Rodella, A.-S., Russ, J., & Zaveri, E. (2017). Uncharted waters: The new economics of water scarcity and variability. In *Uncharted waters: The new economics of water scarcity and variability*. https://doi.org/10.1596/978-1-4648-1179-1

El Haj Assad, M., Ahmadi, M. H., Sadeghzadeh, M., Yassin, A., & Issakhov, A. (2021). Renewable hybrid energy systems using geothermal energy: hybrid solar thermal–geothermal power plant. *International Journal of Low-Carbon Technologies*, *16*(2), 518–530. https://doi.org/10.1093/ijlct/ctaa084

El-Agouz, S. A., El-Samadony, Y. A. F., & Kabeel, A. E. (2015). Performance evaluation of a continuous flow inclined solar still desalination system. *Energy Conversion and Management*, *101*, 606–615. https://doi.org/10.1016/j.enconman.2015.05.069

Elango, T., Kannan, A., & Kalidasa Murugavel, K. (2015). Performance study on single basin single slope solar still with different water nanofluids. *Desalination*, *360*, 45–51. https://doi.org/10.1016/j.desal.2015.01.004

Elfaqih, A. K., Elbaz, A., & Akash, Y. M. (2024). A review of solar photovoltaic-powered water desalination technologies. *Sustainable Water Resources Management*, *10*(3), 1–20. https://doi.org/10.1007/s40899-024-01067-6

Eltawil, M. A., Zhengming, Z., & Yuan, L. (2009). A review of renewable energy technologies integrated with desalination systems. *Renewable and Sustainable Energy Reviews*, *13*(9), 2245–2262. https://doi.org/10.1016/j.rser.2009.06.011

Esmaeilion, F. (2020). Hybrid renewable energy systems for desalination. *Applied Water Science*, *10*(3). https://doi.org/10.1007/s13201-020-1168-5

Eze, J. et al. (2012). Comparative evaluation of rectangular and pyramid-shaped solar stills using saline water. *Academicjournals.Org*. https://academicjournals.org/journal/IJPS/article-full-text-pdf/AA3332116901

Feria-Díaz, J. J., Correa-Mahecha, F., López-Méndez, M. C., Rodríguez-Miranda, J. P., & Barrera-Rojas, J. (2021). Recent desalination technologies by hybridization and integration with reverse osmosis: A review. *Water (Switzerland)*, *13*(10), 1–40. https://doi.org/10.3390/w13101369

Fernandez-Gonzalez, C., Dominguez-Ramos, A., Ibañez, R., & Irabien, A. (2015). Sustainability assessment of electrodialysis powered by photovoltaic solar energy for freshwater production. *Renewable and Sustainable Energy Reviews*, *47*, 604–615. https://doi.org/10.1016/j.rser.2015.03.018

Fthenakis, V., Yetman, G., Zhang, Z., Squires, J., Atia, A. A., Alarcón-Padilla, D. C., Palenzuela, P., Vicraman, V., & Zaragoza, G. (2022). A solar energy desalination analysis tool, sedat, with data and models for selecting technologies and regions. *Scientific Data*, *9*(1), 1–20. https://doi.org/10.1038/s41597-022-01331-4

Gabrielli, P., & Mazzotti, M. (2019). Solar-driven humidification–dehumidification process for water desalination analyzed and optimized via equilibrium theory. *ACS Publications*, *58*(33), 15244–15261. https://doi.org/10.1021/acs.iecr.9b02823

Garg, K., Khullar, V., Das, S. K., & Tyagi, H. (2018). Performance evaluation of a brine-recirculation multistage flash desalination system coupled with nanofluid-based direct absorption solar collector. *Renewable Energy*, *122*, 140–151. https://doi.org/10.1016/j.renene.2018.01.050

Ghaffour, N., Bundschuh, J., Mahmoudi, H., & Goosen, M. F. A. (2015). Renewable energy-driven desalination technologies: A comprehensive review on challenges and potential applications of integrated systems. *Desalination*, *356*, 94–114. https://doi.org/10.1016/j.desal.2014.10.024

Giwa, A., Fath, H., & Hasan, S. (2016). Humidification–dehumidification desalination process driven by photovoltaic thermal energy recovery (PV-HDH) for small-scale sustainable water and power. *Desalination*. https://www.sciencedirect.com/science/article/pii/S0011916415300795

Gnanadason, M. K., Kumar, P. S., Wilson, V. H., & Kumaravel, A. (2015). Productivity enhancement of a-single basin solar still. *Desalination and Water Treatment*, *55*(8), 1998–2008. https://doi.org/10.1080/19443994.2014.930701

Goetzberger, A., Hebling, C., Schock, H. (2003). Photovoltaic materials, history, status and outlook. *Materials Science and Engineering: R: Reports*. https://www.sciencedirect.com/science/article/pii/S0927796X0200092X

Goh, P. S., Matsuura, T., Ismail, A. F., & Ng, B. C. (2017). The Water–Energy Nexus: Solutions towards energy-efficient desalination. *Energy Technology*, *5*(8), 1136–1155. https://doi.org/10.1002/ENTE.201600703

Haddad, Z., Chaker, A., & Rahmani, A. (2017). Improving the basin type solar still performances using a vertical rotating wick. *Desalination*, *418*, 71–78. https://doi.org/10.1016/j.desal.2017.05.030

He, C., Liu, Z., Wu, J., Pan, X., Fang, Z., Li, J., & Bryan, B. A. (2021). Future global urban water scarcity and potential solutions. *Nature Communications*, *12*(1), 1–11. https://doi.org/10.1038/s41467-021-25026-3

Heidary, B., Hashjin, T. T., Ghobadian, B., & Roshandel, R. (2018). Optimal integration of small scale hybrid solar wind RO-MSF desalination system. *Renewable Energy Focus*, *27*(00), 120–134. https://doi.org/10.1016/j.ref.2018.05.003

Hilarydoss, S. (2023). A detailed review on various aspects of inverted solar still desalination systems proposed for clean water production. In *Environmental science and pollution research international* (Vol. *30*, Issue 57). Berlin Heidelberg: Springer. https://doi.org/10.1007/s11356-023-30760-w

Hosseini, M., & Sarhaddi, F. (2017). Performance assessment of a humidification-dehumidification desalination unit connected to photovoltaic thermal collectors. *49*(3), 231–234. https://doi.org/10.22060/mej.2016.765

Houcine, I., Benjemaa, F., Chahbani, M. H., & Maalej, M. (1999). Renewable energy sources for water desalting in Tunisia. *Desalination*, *125*(1–3), 123–132. https://doi.org/10.1016/S0011-9164(99)00130-7

Iaquaniello, G., Salladini, A., Mari, A. et al. (2014). Concentrating solar power (CSP) system integrated with MED–RO hybrid desalination. *Desalination*. https://www.sciencedirect.com/science/article/pii/S0011916413006279

Jackson, K., Jernigan, S., May, W., Maui Oyala, K., & Maui, K. (2022). *Final Report on Humidification-Dehumidification Desalination Prototype*. https://digitalcommons.trinity.edu/engine_designreports/53/

Jones, E., Smakhtin, V., Qadir, M., Van Vliet, M. T. H., Kang, S.-M., & Qadir, M. (2019). The state of desalination and brine production: A global outlook. *Science of the Total Environment*, *657*, 1343–1356. https://doi.org/10.1016/j.scitotenv.2018.12.076

Kabeel, A. E., Arunkumar, T., Denkenberger, D. C., & Sathyamurthy, R. (2017). Performance enhancement of solar still through efficient heat exchange mechanism – A review. *Applied Thermal Engineering*, *114*, 815–836. https://doi.org/10.1016/j.applthermaleng.2016.12.044

Kabeel, A. E., & El-Said, E. M. S. (2013). Technological aspects of advancement in low-capacity solar thermal desalination units. *International Journal of Sustainable Energy*, *32*(5), 315–332. https://doi.org/10.1080/14786451.2012.757613

Kabeel, A. E., Omara, Z. M., Essa, F. A., Abdullah, A. S., Arunkumar, T., & Sathyamurthy, R. (2017). Augmentation of a solar still distillate yield via absorber plate coated with black nanoparticles. *Alexandria Engineering Journal*, *56*(4), 433–438. https://doi.org/10.1016/j.aej.2017.08.014

Kalaivani, S., Radhakrishnan, S. R., Selvakumar, B., & Indhumathy, M. (2012). Experimental Behaviour of a Pyramid Type Solar Still Coupled and Decoupled to an Electrical Temperature Controller. *IJCRR*, *04*(8), 63–76.

Kalbasi, R., Alemrajabi, A., & Afrand, M. (2017). Thermal modeling and analysis of single and double effect solar stills: an experimental validation. *Applied Thermal Engineering*. https://doi.org/10.1016/j.applthermaleng.2017.10.012

Kalogirou, S. A. (2005). Seawater desalination using renewable energy sources. *Progress in Energy and Combustion Science*, *31*(3), 242–281. https://doi.org/10.1016/j.pecs.2005.03.001

Kannakumar, K., Manikandan, P., Ravivarman, G., Sivalingam, A., Moorthy, A. A., Karthik, M., & Girimurugan, R. (2023). Enhanced behavior of solar stills with modified absorber surfaces: Comparative analysis of floating wick designs. *E3s-Conferences.Org*. https://doi.org/10.1051/e3sconf/202345502002

Kapnichor, Y. S. (2017). A Review on Solar Still Water Purification. *International Journal for Innovative Research in Science & Technology*, *3*(09), 59–63.

Kelly, N., & Gibson, T. (2011). Increasing the solar photovoltaic energy capture on sunny and cloudy days. *Solar Energy*. https://doi.org/10.1016/j.solener.2010.10.015

Khan, M. A., Bhuibhar, A. G., & Pande, P. P. (2018). Performance & Analysis and Optimization of Stepped Type Solar Still (a Review). *International Journal of Engineering Sciences & Research Technology*, *7*(3), 69–74.

Kroiß, A., Präbst, A., Hamberger, S. et al. (2014). Development of a seawater-proof hybrid photovoltaic/thermal (PV/T) solar collector. *Energy Procedia*. https://www.sciencedirect.com/science/article/pii/S1876610214009205

Kumar, U. (2017). *Integration of Membrane Distillation and Solar Thermal Systems for Co-production of Purified Water and Heat* (Doctoral dissertation, Ph. D. dissertation, Dept. Eng. Tech., KTH School of Industrial Engineering and Management., English).

Kumar, L., Soomro, J., Khoharo, H., & El Haj Assad, M. (2023). A comprehensive review of solar thermal desalination technologies for freshwater production. *AIMS Energy*, *11*(2), 293–318. https://doi.org/10.3934/ENERGY.2023016

Lawal, D., Antar, M., Khalifa, A., Zubair, S., & Al-Sulaiman, F. (2018). Humidification-dehumidification desalination system operated by a heat pump. *Energy Conversion and Management*, *161*, 128–140. https://doi.org/10.1016/j.enconman.2018.01.067

Leblanc, J. (2009). *Solar Thermal Desalination – A Modelling and Experimental Study*. March, 489. https://doi.org/10.1063/1.2806091

Li, K., Liu, C., Jiang, S., & Chen, Y. (2020). Review on hybrid geothermal and solar power systems. *Journal of Cleaner Production*, *250*. https://doi.org/10.1016/j.jclepro.2019.119481

Mabrouk, A. N., & Fath, H. E. S. (2015). Technoeconomic study of a novel integrated thermal MSF-MED desalination technology. *Desalination*, *371*, 115–125. https://doi.org/10.1016/j.desal.2015.05.025

Mamouri, S., Derami, H., Ghiasi, M., Shafii, M. et al. (2014). Experimental investigation of the effect of using thermosyphon heat pipes and vacuum glass on the performance of solar still. *Energy*. https://www.sciencedirect.com/science/article/pii/S0360544214009414

Manchanda, H., & Kumar, M. (2015). A comprehensive decade review and analysis on designs and performance parameters of passive solar still. In *Renewables: Wind, Water, and Solar* (Vol. *2*, Issue 1). Wind, Water, and Solar, no. 1, Springer Singapore. https://doi.org/10.1186/s40807-015-0019-8

Manju, S., & Sagar, N. (2017). Renewable energy integrated desalination: A sustainable solution to overcome future fresh-water scarcity in India. *Renewable and Sustainable Energy Reviews*, *73*, 594–609. https://doi.org/10.1016/j.rser.2017.01.164

Manolakos, D., Kosmadakis, G., Kyritsis, S., & Papadakis, G. (2009). On site experimental evaluation of a low-temperature solar organic Rankine cycle system for RO desalination. *Solar Energy*, *83*(5), 646–656. https://doi.org/10.1016/j.solener.2008.10.014

Mayere, A. (2011). Solar powered desalination. PhD thesis, University of Nottingham. *Book*, *1*(1).

Meratizaman, M., & Godarzi, A. A. (2019). Techno-economic feasibility of a solar-powered reverse osmosis desalination system integrated with lithium battery energy storage. *Desalination and Water Treatment*, *167*, 57–74. https://doi.org/10.5004/dwt.2019.24550

Mohammadi, K., Saghafifar, M., Ellingwood, K., & Powell, K. (2019). Hybrid concentrated solar power (CSP)-desalination systems: A review. In *Desalination* (Vol. *468*). https://doi.org/10.1016/j.desal.2019.114083

Mohammed, R. H., & Askalany, A. A. (2019). Productivity improvements of adsorption desalination systems. In *Green Energy and Technology* (pp. 325–357). https://doi.org/10.1007/978-981-13-6887-5_15

Morin, O. J. (1993). Design and operating comparison of MSF and MED systems. *Desalination*, *93*(1–3), 69–109. https://doi.org/10.1016/0011-9164(93)80097-7

Nafey, A. S., Fath, H. E. S., El-Helaby, S. O., & Soliman, A. (2004). Solar desalination using humidification-dehumidification processes. Part II. An experimental investigation. *Energy Conversion and Management*, *45*(7–8), 1263–1277. https://doi.org/10.1016/S0196-8904(03)00152-3

Nafey, A. S., Fath, H. E. S., & Mabrouk, A. A. (2006). Thermo-economic investigation of multi effect evaporation (MEE) and hybrid multi effect evaporation-multi stage flash (MEE-MSF) systems. *Desalination*, *201*(1–3), 241–254. https://doi.org/10.1016/j.desal.2005.09.044

Naroei, M., Sarhaddi, F., & Sobhnamayan, F. (2018). Efficiency of a photovoltaic thermal stepped solar still: Experimental and numerical analysis. *Desalination*. https://www.sciencedirect.com/science/article/pii/S0011916417312535

Orfi, J., Laplante, M., Marmouch, H., Galanis, N., Benhamou, B., Nasrallah, S. B., & Nguyen, C. T. (2004). Experimental and theoretical study of a humidification-dehumidification water desalination system using solar energy. *Desalination*, *168*(1–3), 151–159. https://doi.org/10.1016/j.desal.2004.06.181

Panchal, H., Sadasivuni, K. K., Israr, M., & Thakar, N. (2019). Various techniques to enhance distillate output of tubular solar still: A review. In *Groundwater for Sustainable Development* (Vol. *9*). https://doi.org/10.1016/j.gsd.2019.100268

Panchal, H. N., & Shah, P. K. (2012). Performance Improvement of Solar Stills via Experimental Investigation. *International Journal of Advanced Design and Manufacturing Technology*, *5/No. 5*(5), 19–23.

Parekh, S., Farid, M. M., Selman, J. R., & Al-Hallaj, S. (2004). Solar desalination with a humidification-dehumidification technique – A comprehensive technical review. *Desalination*, *160*(2), 167–186. https://doi.org/10.1016/S0011-9164(04)90007-0

Pourkiaei, S. M., Ahmadi, M. H., Ghazvini, M., Moosavi, S., Pourfayaz, F., Kumar, R., & Chen, L. (2021). Status of direct and indirect solar desalination methods: comprehensive review. In *European Physical Journal Plus* (Vol. *136*, Issue 5). Berlin Heidelberg: Springer. https://doi.org/10.1140/epjp/s13360-021-01560-3

Pouyfaucon, A. B., & García-Rodríguez, L. (2018). Solar thermal-powered desalination: A viable solution for a potential market. In *Desalination* (Vol. *435*, pp. 60–69). https://doi.org/10.1016/j.desal.2017.12.025

Rajaseenivasan, T., & Srithar, K. (2016). Performance investigation on solar still with circular and square fins in basin with CO2 mitigation and economic analysis. *Desalination*. https://www.sciencedirect.com/science/article/pii/S0011916415301272

Raval, H., & Maiti, S. (2015). A novel photovoltaic powered reverse osmosis with improved productivity of reverse osmosis and photovoltaic panel. *Journal of Membrane Science & Research*, *1*, 113–117. https://www.msrjournal.com/article_14481.html

Raval, H., & Maiti, S. (2016). Ultra-low energy reverse osmosis with thermal energy recovery from photovoltaic panel cooling and TFC RO membrane modification. *Desalination and Water Treatment*. https://www.tandfonline.com/doi/abs/10.1080/19443994.2014.993725

Reif, J. H., & Alhalabi, W. (2015). Solar-thermal powered desalination: Its significant challenges and potential. *Renewable and Sustainable Energy Reviews*, *48*, 152–165. https://doi.org/10.1016/j.rser.2015.03.065

Rognoni, M. (2007). Solar desalination. *International Journal of Nuclear Desalination*, 2(4), 363–371. https://doi.org/10.1504/IJND.2007.015802

Sampathkumar, K., Arjunan, T. V., Pitchandi, P., & Senthilkumar, P. (2010). Active solar distillation-A detailed review. *Renewable and Sustainable Energy Reviews*, *14*(6), 1503–1526. https://doi.org/10.1016/j.rser.2010.01.023

Sansom, C., & Tonnellier, X. (2018). *A concentrated solar still for community scale desalination*. https://dspace.lib.cranfield.ac.uk/handle/1826/19378

Saxena, A., Agarwal, N., & Srivastava, G. (2013). Design and performance of a solar air heater with long term heat storage. *International Journal of Heat and Mass Transfer*. https://www.sciencedirect.com/science/article/pii/S0017931012010009

Schallenberg-Rodríguez, J., Del Rio-Gamero, B., Melian-Martel, N., Lis Alecio, T., & González Herrera, J. (2020). Energy supply of a large size desalination plant using wave energy. Practical case: North of Gran Canaria. *Applied Energy*, *278*, 115681. https://doi.org/10.1016/J.APENERGY.2020.115681

Sharaf, M. A., Nafey, A. S., & García-Rodríguez, L. (2011). Exergy and thermo-economic analyses of a combined solar organic cycle with multi effect distillation (MED) desalination process. *Desalination*, *272*(1–3), 135–147. https://doi.org/10.1016/j.desal.2011.01.006

Sharon, H., & Reddy, K. S. (2015). A review of solar energy driven desalination technologies. *Renewable and Sustainable Energy Reviews, 41*, 1080–1118. https://doi.org/10.1016/j.rser.2014.09.002

Shrivastava, A., Rosenberg, S., Peery, M. (2015). Energy efficiency breakdown of reverse osmosis and its implications on future innovation roadmap for desalination. *Desalination*. https://www.sciencedirect.com/science/article/pii/S0011916415000090

Sivaku, R., & Jidhesh, P. (2016). Improvement techniques in performance and productivity of solar stills. *International Journal of Information Science and Computing, 6*(9), 2441–2450.

Suneesh, P., Jayaprakash, R., Kumar, S., & Denkenberg, D. (2017). Performance analysis of "V"-type solar still with tilt wick and effect of wick coverage. *Taylor & Francis, 4*(1), 1419791. https://doi.org/10.1080/23311916.2017.1419791

Tiwari, A., & Tiwari, G. N. (2006). Effect of water depths on heat and mass transfer in a passive solar still: in summer climatic condition. *Desalination*. https://www.sciencedirect.com/science/article/pii/S0011916406003997

Tiwari, A., & Kumar, A. (2023). A comprehensive review on solar thermal desalination systems based on humidification-dehumidification approach. *Clean Technologies and Environmental Policy, 25*(7), 2075–2097. https://doi.org/10.1007/s10098-023-02517-z

Tiwari, G., Yadav, Y., Eames, P. & Norton, B. (1994). Solar distillation systems: The state-of-the-art in design development and performance analysis. *Renewable Energy*. https://www.sciencedirect.com/science/article/pii/0960148194904251

Ullah, I., & Rasul, M. G. (2019). Recent developments in solar thermal desalination technologies: A review. *Energies, 12*(1). https://doi.org/10.3390/en12010119

Velliangiri, M., Sengottain, S., Balasundaram, J., Manikandan, V., Shanmugasundaram, K., Shanmugan, S., Janarthanan, B., & Chandrasekaran, J. (2013). Wick type solar stills: A review. *Renewable and Sustainable Energy Reviews*. https://doi.org/10.1016/j.rser.2012.11.046

Vigil, K. M. (2023). *Clean Water, 2nd ed: An Introduction to Water Quality and Water Pollution Control*. https://doi.org/10.1353/book83960

Vishwanath Kumar, P., Kumar, A., Prakash, O., & Kaviti, A. K. (2015). Solar stills system design: A review. *Renewable and Sustainable Energy Reviews, 51*, 153–181. https://doi.org/10.1016/j.rser.2015.04.103

Yadav, S., & Sudhakar, K. (2015). Different domestic designs of solar stills: A review. *Renewable and Sustainable Energy Reviews, 47*, 718–731. https://doi.org/10.1016/j.rser.2015.03.064

Yuan, G., Wang, Z., Li, H., & Li, X. (2011). Experimental study of a solar desalination system based on humidification–dehumidification process. *Desalination*. https://www.sciencedirect.com/science/article/pii/S0011916411003183

Zuo, L., Chen, J., Zhou, X., & Xu, B. (2019). A preliminary study of wind supercharged solar chimney power plant combined with seawater desalination. *Taiyangneng Xuebao/Acta Energiae Solaris Sinica, 40*(3), 620–627. https://www.sciencedirect.com/science/article/pii/S0038092X1830272X

Wind Energy Integrated Desalination

5

5.1 INTRODUCTION

Energy and water are interconnected resources crucial in establishing our existence and advancing human society (Serrano-Tovar et al., 2019). The well-being and prosperity of modern society rely on the sustainable provision of water and energy. Currently, around three billion individuals lack access to a reliable and safe supply of freshwater, while around 1.76 billion people reside in regions already experiencing significant water scarcity (Rashidi et al., 2022). As the population grows and industrial and agricultural activity rises, water supplies have been extensively exploited and polluted. The demand for potable water is a vital issue on the global agenda, ranking equally high in importance as climate change (Serrano-Tovar et al., 2019).

The continuous global warming, population growth, urbanization, and rising water and energy use consistently damage our invaluable biosphere (Dahioui & Loudiyi, 2013). These unsustainable practices are perpetuated by commercial and governmental structures that favor the growth of water-intensive industries in arid and semiarid areas with scarce water resources (Damania et al., 2017). This is also a prevalent phenomenon in developing nations, particularly in regions like Africa and Asia, where there is a critical scarcity of water (Jarquin-Laguna & Greco, 2019). The current's environmental, social, and economic well-being relies on a sustainable provision of water and energy. Many places have realized that current freshwater supplies are insufficient to meet the increasing needs, and in certain circumstances, there is a lack of suitable groundwater and surface water sources. Hence, in recent decades, the necessity of harnessing saline water from the seas and oceans and the methods for transforming salt water into freshwater have been acknowledged as rational strategies (Hossam-Eldin et al., 2017).

Initially, desalination systems were expensive and required a significant amount of energy. The motivation to implement desalination plants in numerous coastal and urban cities to meet freshwater demands arises from (a) advancements in energy efficiency of

DOI: 10.1201/9781003530893-5

various desalination technologies and (b) decreased capital expenditure for desalination procedures (Kumar et al., 2023). Although desalination of saline waters is increasingly recognized as a viable option for obtaining sources of freshwater, the energy requirements of current desalination methods remain a significant obstacle to their practical use (Novosel et al., 2014). Studies have determined that producing 1000 m^3 of freshwater daily requires 10,000 m^3 of petroleum annually. Conventional fossil fuel sources are not sustainable energy sources (Novosel et al., 2014). Utilizing energy from environmentally friendly sources, like wind, solar, and geothermal, presents a feasible and sustainable alternative (Bundschuh & Hoinkis, 2019).

Wind energy, a highly efficient renewable energy (RE), can potentially reduce the adverse effects of seawater desalination on the environment by 75%. Furthermore, wind energy has a very low water consumption throughout its life cycle (Ma & Lu, 2011). Various technological devices have been created to harness the kinetic energy of air. The primary objective has usually been to harness the kinetic energy produced by moving air masses through revolving wind turbines (WTs) (Heidary et al., 2018). This energy is subsequently transformed into mechanical energy, which is utilized to rotate a shaft linked to a capturing apparatus called the rotor. Mechanical energy can be directly employed or transformed into alternative energy types for many applications. Rapid WTs are the predominant kind used for generating power due to their versatile application and increased interest. Most of the time, people use the word "wind turbines" in this case (Saravana & Rajini, 2014). One of the most well-known types of vertical-axis engines is the Darrieus, named after the French engineer who invented it. It was made in the year 1931. Right now, the three-bladed horizontal-axis form is the most common type of WT.

WTs have substantially grown in size during the past 25 years (Koklas & Papathanassiou, 2006). A notable recent trend has been the emergence of commercially available WTs with power capacities ranging in the megawatt range. These arrangements are utilized in offshore and onshore wind farms to transfer their electrical energy into traditional power systems (Figure 5.1) (Mersin & Çeliktaş, 2023).

Figure 5.1 Diagram illustrating the classification of wind turbines by their rated power: small (50–100 kW, 15–20 m), medium (500 kW, 40 m), large (200–5000 kW, 80–124 m), displayed with increasing height and capacity.

With recently built wind farms offering energy pricing comparable with traditional electricity-generating systems, wind technology has emerged as one of the most cost-effective, sustainable options (Lotfy et al., 2022). Wind power can be utilized as a cheap fuel in many areas, even without government subsidies. However, the exact cost of the energy generated by a wind farm mainly depends on the local wind resource and the project's financing method. If water desalination is also necessary in a region with a high wind potential, wind power is the way to go (Ghaffour et al., 2015).

The primary objective of most wind energy applications in integration with seawater reverse osmosis desalination (SWRO) is to utilize traditional wind technology to generate the necessary electricity for seawater desalination operations (Forstmeier et al., 2007). WTs first transform the kinetic energy from wind movement into electrical and mechanical energy (Liu et al., 2002). Subsequently, the electricity from the grid is

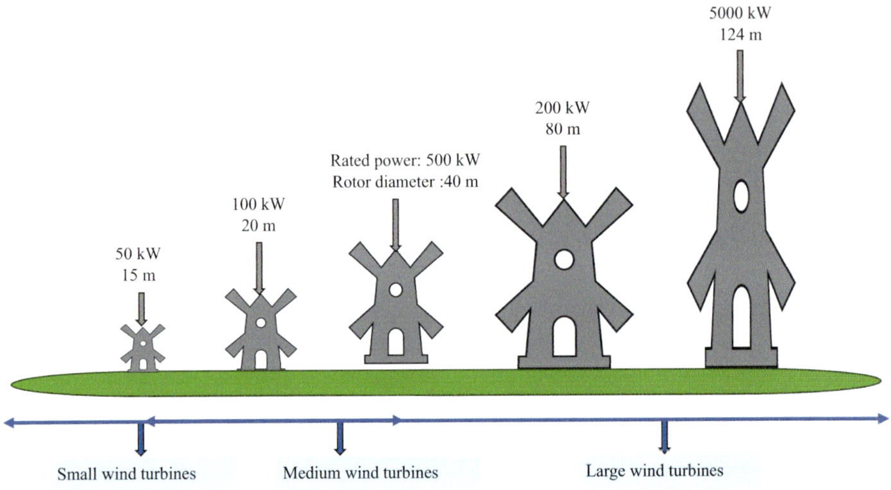

FIGURE 5.1 Classification of wind turbines according to their rated power.

transformed into the mechanical power necessary for the pumps to operate the SWRO process, consuming around 4 to 6 kWh/m^3 of freshwater generated (Weiner et al., 2001). The current technique comprises many energy conversion phases, utilizing the grid as an energy buffer to accommodate the changes in wind output (Cabrera et al., 2021; García-Rodríguez, 2004).

This chapter explores integrating wind energy with desalination technologies to address the critical global issues of water scarcity and sustainable energy. It highlights the interconnection between energy and water resources, emphasizing the need for innovative solutions to provide freshwater in regions facing significant shortages. By examining various case studies and technological approaches, the chapter aims to demonstrate the feasibility and economic viability of using wind energy to power desalination units, ultimately contributing to a sustainable and reliable freshwater supply.

Section 5.2 discusses the unique traits of wind energy compared to traditional energy sources. The section emphasizes the importance of accurate wind data measurement and site evaluation to optimize WT performance. It also highlights the need for continuous wind monitoring and meteorological assessments to ensure the feasibility and efficiency of wind energy projects. Section 5.3 discusses the integration of wind power with desalination technologies. It highlights how wind energy can power desalination systems, particularly in isolated regions like islands, where water production costs can be reduced. The section explores different methodologies for connecting WTs to desalination systems, including direct connections and grid-connected systems. It also addresses the challenges of power fluctuations and the need for supplementary systems like batteries or diesel generators. Section 5.4 discusses using wind power for desalination in both off-grid and on-grid scenarios. The section highlights the importance of strategically placing WTs, energy storage, and integrating renewable energy sources (RES) to meet water demands efficiently. It also emphasizes specialized energy management units for on-grid and off-grid applications. The case studies section highlights various implementations of wind-powered desalination systems globally.

5.2 WIND ENERGY CHARACTERISTICS AND EXPLOITATION

Wind energy possesses traits opposite to traditional energy sources' defining attributes (García-Rodríguez, 2004). The most prominent advantages of wind energy over conventional energy sources are its renewable and non-polluting characteristics. Moreover, wind energy is extensively accessible in several regions throughout. However, wind energy has specific attributes that prevent it from experiencing several inherent advantages to traditional fossil fuels (Dahioui & Loudiyi, 2013). It is essential to acknowledge that wind energy is characterized by its inherent unpredictability and fluctuation in terms of time and location. Furthermore, unlike traditional energy sources that rely on easily storable and transportable fuels, wind energy cannot be stored in its raw form. Therefore, WTs must be strategically placed with sufficient wind resources (Darwish, 2021).

WT energy curves show the average power outputs as a function of hub height and wind speed for various rotor power controls (Giwa, 2015). The power curve the manufacturer provides for a specific air density can be deceptive for several reasons. These include, but are not limited to, the location of the installation site's orography and roughness, the air density, the number of nearby WTs, and the severity of any turbulence that may be present (Greco et al., 2021).

Creating a wind data collection plan to minimize uncertainty in evaluating the wind resources for the proposed windmills at the location is crucial (Ghaffour et al., 2013). An appropriate quantity of meteorological structures must be constructed to get the necessary data, and these towers should ideally be tall enough to measure the wind properties at the height of the windmill's hub. It is advisable to continuously monitor wind for a minimum of 12 months to determine the seasonal fluctuations in wind characteristics (Ma & Lu, 2011).

The evaluation procedure starts after a possible location for wind exploitation is located using wind maps and other relevant data such as land cover, topography, accessibility, use limitations, existing electricity transmission lines, and environmental impact assessment (Ma & Lu, 2011). The procedure begins by employing data loggers, sensors, and data storage devices to capture weather forecasting parameters, including wind direction and speed, temperature, barometric air pressure, and relative humidity (RH). These factors are then examined to comprehend the wind resource's behavior at the height of the WT's rotor, which is planned to be installed in the region. The analysis process also involves assessing the power production of WTs or wind farms and conducting research to determine their financial and economic feasibility (Lai et al., 2016; Rodríguez et al., 1996).

5.3 WIND ENERGY FOR DESALINATION TECHNOLOGIES

Desalination systems and wind power are two distinct technologies that can be integrated differently. When the energy from the WTs reaches the desalination plant, the two systems connect. This interaction happens at the exact location or area where the energy is transmitted (Novosel et al., 2014).

The desalination of seawater and freshwater production might be powered by wind energy systems, which would benefit remote areas such as coastlines (Dahioui & Loudiyi, 2013). Utilizing these technologies results in a more cost-effective method of producing water than the expenses associated with transporting water to remote islands or burning conventional fuels to create energy. Wind desalination systems can make use of a variety of approaches (García-Rodríguez, 2004). A grid system connects the desalination plant to the WTs. Finding the optimal dimensions of the windmill and desalination systems and estimating the financial savings from not using fuel are the main points of this case study (Mbarga et al., 2014).

WTs can also be directly connected to desalination systems. The desalination process is vulnerable to power variations and interruptions from the wind energy source (Dahioui & Loudiyi, 2013). However, these power fluctuations adversely affect the efficiency and lifespan of specific desalination apparatus (Eltawil et al., 2009). Consequently, auxiliary components such as diesel generators, batteries, or flywheels may be integrated into the system. The primary study goal in this domain entails examining the wind facility and assessing the system's overall performance. Furthermore, endeavors are undertaken to formulate appropriate control techniques for the WTs and the entire system (Lotfy et al., 2022).

Powering various desalination processes, such as electrodialysis (ED) (Oré et al., 2022), reverse osmosis (RO), and mechanical vapor compression (MVC) (Karameldin et al., 2003), requires converting wind energy into electricity (Dahioui & Loudiyi, 2013). Wind power facilities can operate connected to or disconnected from the primary power grid. To address the intermittent operation of wind energy, auxiliary devices such as batteries, water reservoirs, or flywheel mechanisms should be incorporated into the systems. These facilities store or discharge energy when the wind velocity surpasses or falls short of the specified threshold (García-Rodríguez, 2003).

A hybrid reverse osmosis-multi-stage flash (RO-MSF) plant requires substantial energy, including thermal and electrical power (Ghaffour & Mujtaba, 2018). Both energy sources are derived from fossil fuels, leading to air pollution and the emission of greenhouse gases. The hybrid RO-MSF system necessitates access to an electrical power supply and oil or gas, rendering it unsuitable for remote locations (Heidary et al., 2018). Solar and wind energy are the principal renewable sources that meet energy demands in numerous places globally (Ghaffour & Mujtaba, 2018). It is possible to harvest solar radiation and wind energy in Iran's coastal regions because of the favorable conditions that exist there. The availability of freshwater resources in Iran is currently in short supply. A hybrid solar-wind system may optimize the RO-MSF desalination process, utilizing clean RES like wind and solar power, potentially offering an environmentally sustainable alternative for water production (Bundschuh et al., 2021; Manju & Sagar, 2017). Hybridization, which is the process of combining renewable and clean sources of energy such as wind power and solar power, has the potential to improve the accessibility and sustainability of RES while simultaneously reducing the costs of energy (Tzen & Morris, 2003).

The economy and feasibility studies of wind-powered RO systems are becoming more critical to aid in layout, water production, site selection, cost/price calculation, and other related matters due to the increasing crisis in traditional energy sources and the rapid growth of wind energy and RO technologies (Heidary et al., 2018). Koklas and Papathanassiou (2006) presented a logistic model to shed light on the factors to be considered while choosing components for a standalone wind-driven RO system. Several different configurations were used to simulate the system's operation. These configurations altered the system's key components' sizes, including the WT, the RO plant, and the batteries. The cost of producing water was analyzed for each of the many scenarios that were included in the economic evaluation that was carried out.

Currently, the majority of desalination plants throughout the world are powered by fossil fuels. Wind-driven power production projects have been successfully incorporated into several modern SWRO desalination plants in Australia. These projects provide electricity equivalent to the power the desalination plants require. To supply electricity for saltwater desalination, several countries in North Africa and the Middle East have recently made efforts to create a robust collection of renewable power-producing facilities. Under 1% of the total desalination capacity in the North Africa and Middle East region is currently being utilized by RES for desalination (Elimelech & Phillip, 2011). The MENA region's growing dependence on desalinated water and the expensive production costs and substantial carbon emissions associated with fossil fuel-powered desalination renders it unsustainable in the long term (Maftouh et al., 2023).

Koroneos et al. (2007) established an integrated model within the REDDES program to utilize renewable energies (wind, solar) for seawater desalination. The desalination technologies, specifically reverse osmosis (RO) and MVC, were combined with RES to produce drinkable water at the most affordable cost. The findings demonstrated that the expenses associated with producing water in an RES-desalination setup were significantly influenced by the extent of the accessible RES capacity. As the potential for RE increases, the cost of energy production from RES units decreases, resulting in lower costs for water production from desalination units.

Feron initially suggested the idea of a safe operating window for the temporary running of a seawater RO system powered by wind. Based on the findings, operating the plant from an intermittent and fluctuating power supply is safe, provided that the on/off cycling is managed to avoid damaging the membrane due to the rate of pressure change and cycling frequency (GE Global Research, 2004). A curved-sided quadrilateral safe operating window (SOW) was derived by Feron using the restrictions of the membrane properties, which include maximum concentrate flow rate, minimum concentrate flow rate, maximum feed water pressure, and maximum permeate concentration. First, he suggested adjusting the membrane area; second, he suggested enabling a transient operation within the safe operating window limits so that wind-membrane seawater plants could handle a changing wind resource (Ghaffour et al., 2013). Although he acknowledged the increasing complexity of the plant, he determined that both recommendations were economically unsustainable in 1995 due to the underutilization of an expensive membrane area and the comparatively little gain in production.

Forstmeier et al. (2007) created a model to validate the technological viability of utilizing wind as the energy supply for desalination, encompassing RO and MVC units. The findings show that the expenses aligned with the anticipated amount for a traditional desalination system, establishing its significant cost competitiveness in regions abundant in wind resources and burdened by high energy costs.

Rodríguez et al. (1996) examined the impact of critical factors on freshwater costs, including design, operating conditions, climatic conditions, salt concentration, WT power, plant capacity, RO module cost, and WT cost. A wind-powered RO system's production cost was determined by Romero-Ternero et al. (2005). Using thermoeconomy, they also calculated the process's exergy efficiency. Only by testing a wind-powered desalination plant was the price of freshwater per unit determined. The comprehensive economic analysis determined that wind-powered saltwater reverse osmosis desalination was the most economical choice for that specific wind energy site. The desalination plant's capacity was moderate, and the cost of producing 1 m^3 of freshwater was 76 cents.

Liu et al. (2002) conducted experiments on a wind-powered RO system employing produced freshwater. The system lacked energy storage and was evaluated under diverse wind speeds in varying conditions. At wind speeds of 7.0 m/s or higher, both the permeate flux and salinity (in terms of quantity and quality) remained stable despite turbulence intensities approaching 0.4. This was seen within the designated operating range and was unaffected by the duration of the variations. Despite significant power changes, the researchers determined that their wind-membrane system could function within a safe operating window.

Koutroulis and Kolokotsa (2010) created a method for determining the proper measurements of photovoltaic (PV) modules and wind-driven RO systems. The appropriate quantity and type of system elements were identified from various commercially available options. The objective was to decrease the overall system cost over 20 years while satisfying consumer water demand. The ideal sizing results demonstrated that the operational characteristics of its components considerably influenced the total cost of the RO system.

An innovative experimental investigation on the direct connection of a system powered by winds and an RO facility using shaft power was carried out at the Canary Islands Technological Institute as part of projects AERODESA I and AERODESA II. The study was undertaken between the years of 2010 and 2012. In addition, an investigation was carried out on an RO plant powered by wind and removing salt from brackish water. This facility is located on Coconut Island, close to the northern coast of Oahu, Hawaii. The high-pressure pump is instantly connected to the power that is generated by the windmill's shaft by the system. This connection makes it possible to maintain a flow rate of 13 L/min even when the wind speed is 5 m/s (Eltawil et al., 2008).

In 2002, Soslaires Canarias S.L. built a SWRO desalination facility near Playa de Vargas on the east coast of Gran Canaria island. Additionally, it comprises the activities of water pumping, the installation of pipelines, and the construction of a water reservoir. To exceed the initial water demands of the crops farmed in this region, the desalination plant was developed with a capacity of 5000 m^3/day. In addition to providing power to the plant and water pumping activities, the electrical grid is also connected to a wind power farm located on land with a capacity of 2.64 megawatts. If wind energy is not sufficient, the SWRO system will pull electricity from the grid. An additional investment bonus of 12 euros/MW is added to the market price of the extra power produced by the wind farm, which is then sold to the transmission grid (Rybar et al., 2005).

5.4 OFF-GRID AND ON-GRID WIND POWER FOR DESALINATION

Wind power systems "connected to the grid" are already in use in some places. RE technology must provide the desalination units with the power they need through a power node connected to the grid (Zheng & Qian, 2019). This way, there is no need for any other power backup sources. Suppose there is an excess of energy produced by RES. In that case, this surplus is supplied to the grid, and the producer receives a feed-in tariff, subject to the country's procedures and legislation. These incidents mostly pertain to using a WT to fulfill the power needs of RO plants (Zuo et al., 2019).

The term "off-grid" refers to initiatives carried out in distant and rural locations when there is no access to the power grid, and there are issues with water scarcity (Francisco et al., 2018). It is possible to employ minimal independent wind desalination systems to meet the water and fundamental power supply requirements in several rural locations worldwide. This is because the population is dispersed in these places. The low cost, low maintenance needs, easy operation, high reliability, and sustainability are the five most desirable characteristics for such systems. Other desirable characteristics include the low cost (Zuo et al., 2019).

Most on-grid applications include larger units, in which a wind farm or a WT can meet the energy needs of the desalination units. The application of this method is limited to regions with access to the electrical wiring grid. These wind desalination technologies have several advantages; the desalination unit may continue to function dependably and continuously. The wind farm or WT can sell any excess energy it generates back to the grid of RES. Additionally, the system's monitoring control and total automation are significant features (Cabrera et al., 2021). It is recommended that a specialized energy management system be developed for both on-grid and off-grid systems separately. When designing and installing a WT, three of the most important technical characteristics are the wind potential and the siting concerns (Koklas & Papathanassiou, 2006).

The availability of electricity, the capacity to store energy and water, and the demand for potable water should all be considered when determining the desalination system's capacity. It is not advisable to satisfy the water needs in off-grid locations by selecting advanced wind desalination couplings. An essential component affecting the everyday operation and life of the system is its simplicity and level of automation (Dahioui & Loudiyi, 2013; García-Rodríguez, 2004). The following are some broad categories of factors that should be addressed while planning and executing a wind desalination project (Figure 5.2).

Figure 5.2 Diagram of a pyramid with nine horizontal steps, outlining the key factors for planning and executing a wind desalination system: wind assessment, water resource assessment, land acquisition, municipal consultations, social acceptance, employment creation, wind farm and desalination design, economic analysis, site preparation, and construction.

Several factors must be considered when designing wind desalination systems in off-grid systems. For instance, the amount of necessary generated water is an essential measurement (Ma & Lu, 2011). Conventional desalination systems typically run for around 24 h/day, and their nominal water capacity is calculated based on that amount of time. For example, an RO machine that operates around the clock can produce 20 m^3 of water daily. The off-grid independent system provides eight hours of electricity daily so that the RO unit can produce approximately 6.6 m^3. The designer must evaluate various

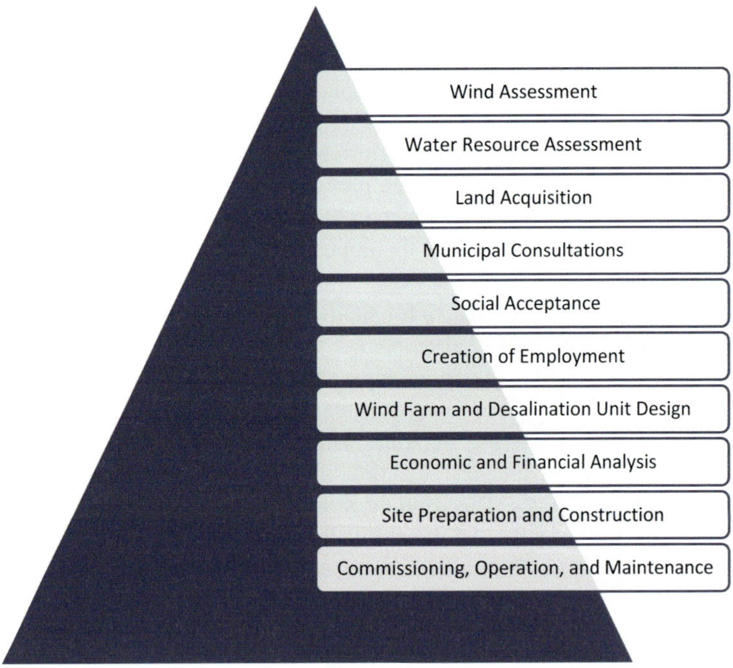

FIGURE 5.2 Factors for planning and executing a wind desalination system.

alternatives, including a substantial RO unit, an expanded power supply system, an enhanced storage system, or a backup system, all of which depend on the allocated budget (Heidary et al., 2018). An energy recovery device (ERD) may allow a bigger capacity RO unit to utilize less energy than a smaller one. Take the 20 m^3/day SWRO device as an example; using ERD, its specific energy consumption is calculated to be around 4.8 kWh/m^3. An increased SWRO with a 100 m^3/day water capacity and ERD is anticipated to have a specific energy consumption of approximately 3.6 kWh/m^3 (Greco et al., 2021; Rashidi et al., 2022).

Wind farms linked to the grid provide enough electricity for numerous desalination facilities in Australia. Kwinana, located south of Perth in Western Australia, is home to a massive SWRO plant built in 2006. During a single day, it can hold 143,000 m^3 of water. The Perth plant was the first primary RO plant in the world, and it made history by obtaining all of its power from renewable sources. Because of ERD, the RO plant has a specific energy consumption of only 3.5 KWh/m^3, which decreases significantly. From the Emu Downs Wind Farm, which is around 200 km north of Perth, the facility receives its electricity from the power generator. This wind farm's 80 MW production capacity is more significant than three times the desalination plant's maximum consumption. With an annual output of about 272 GWh, the 48 WTs compensate entirely for the 180 GWh expected electrical demand of the Perth SWRO facility. There are 48 Vestas (V82) WTs in Emu Downs Wind Farm, and their combined nominal or generating capacity is 1.65 MW (Goosen et al., 2024; Ma & Lu, 2011).

A wind power plant is coupled to the electrical grid on a Greek island in the Cyclades region. A seawater reverse osmosis (SWRO) system that produces 3,600 m^3 of water daily can meet most of Milos Island's water requirements (Greco et al., 2021). A WT with a capacity of 850 kW (VESTAS-V52) connected to the nation's electrical grid provides power to this establishment. "Vouno Triovassalou" on the island of Milos is the location of the RO facility. This facility was formerly associated with the Betonies Mine. The following are the components that make up the project. Daily, the RO unit can produce 3600 m^3 of water, comparable to six units of 600 m^3 each day. It is designed to desalinate saltwater. The project's constituent parts are as follows. The RO unit can generate 3600 m^3 of water per day, which is equivalent to six 600 m^3 units per day. Saltwater desalination is its intended use. Drinkable water can be stored in 3000 m^3 storage tanks (Karytsas et al., 2004). To move drinkable water from the RO unit to the storage tanks, set up a pipeline system. The WT has a power output of 850 kW and is situated in a pre-existing wind park with three turbines. The WT is expected to generate an average of 1,800,000 kWh of electricity annually. In order to keep the units running smoothly in response to the electrical load on Milos Island and the operational status of the thermal plant regulated by the Public Power Corporation (PPC), supervisory control and data acquisition systems (SCADA) are utilized. A RO system with ERDs and high-efficiency pumps can achieve a specific energy consumption of about 2.5 kWh/m^3 (Fytikas et al., 2005; Petrovic et al., 2018).

Figure 5.3 Flow diagram illustrating electricity generation from a wind turbine and a conventional power plant, both linked to grid-connected desalination systems. Power is efficiently distributed to a solar panel and a desalination plant with mechanical and thermal units.

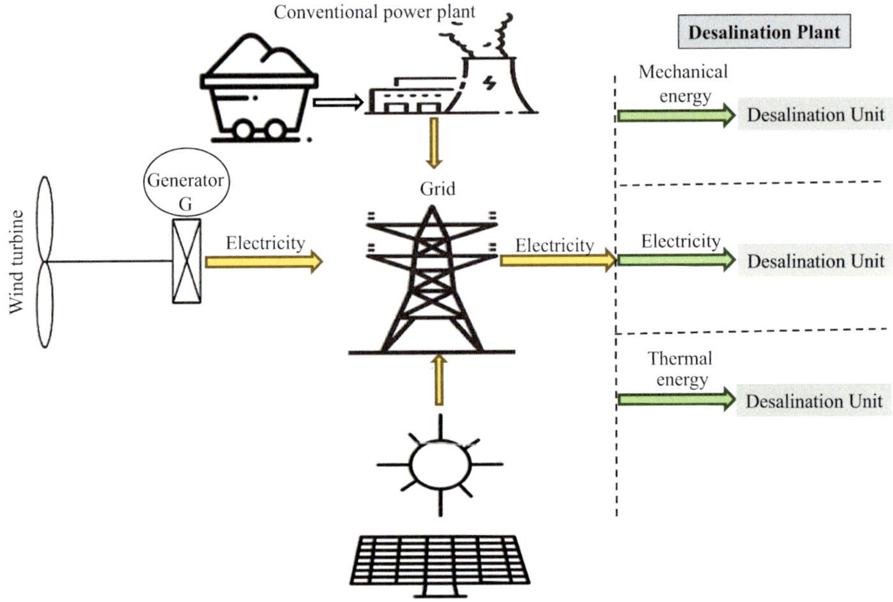

FIGURE 5.3 Grid-connected desalination systems.

5.5 STUDIES ON WIND INTEGRATED DESALINATION SYSTEMS

In 1982, a small system was established on the island of Ile du Planier in France. A turbine with a capacity of 4 kW and a production rate of 0.5 m^3 per hour was contained within the desalination apparatus. In addition, the technology was developed to operate with batteries. Within the context of the Island of Drence, the combination of wind energy and RO was implemented in 1990. To power a seawater RO machine, a WT with a generation capacity of 10 kW was deployed. At a testing center in Lastours, an intriguing experience was obtained. There, a WT with a capacity of 5 kW gave electricity to several batteries with a capacity of 1500 Ah and 24 V. Additionally, the energy was delivered to a RO unit with a rated output of 1.8 kW utilizing an inverter (Eltawil et al., 2008; Ma & Lu, 2011).

In 1991, a pilot plant was built on Germany's Borkum Island. The facility had a 45-kW windmill connected to a 48-m^3/d MVC evaporator. It was possible to build a generator that had 36 kW of power. Since 1995, a bigger factory has been built on Rügen Island in the Baltic Sea. The building could handle 360 m^3/d of material and generate 300 kW of electricity from wind (Ma & Lu, 2011).

A wind-powered MVC plant with a capacity of 50 m^3/day was built in Gran Canaria, Spain, by the Instituto Tecnologico de Canarias (ITC). Sea Desalination Independent Wind Energy System (SDAWES) was a proposal that included this plant as one of its parts. During the construction of the wind farm, there were two WTs with a capacity of 230 kW each, a flywheel that rotated at a speed of 1500 revolutions per minute and was connected to a 100 kVA synchronous machinery, an isolation transformer that was situated in a specific construction, and a 7.5 kW uninterruptible power source that was put in the control dome (Ghaffour et al., 2015). Karameldin et al. (2003) conducted an in-depth investigation of the influence of essential elements in wind-powered MVC systems. According to the findings of the investigation, the temperature of the operational evaporator that was suggested was 50°C, and the temperature differential that was recommended was 3°C. In these locations, the system productivity was 203, 398, and 938 m^3/d, respectively, with windmill dimensions of 20, 28, and 43 meters. This was determined by the wind speeds that were prevalent at the time.

A hybrid wind and PV-driven RO desalination facility was developed by Kershman et al. (2003) near the Mediterranean Sea in Libya. The plant was intended to provide 200 m^3/day of drinkable water for a surrounding community. It was anticipated that the RO desalination machine would require a nominal power of 70 kW (net power post-recovery) to function correctly. The solar PV plant was engineered for a capacity of 50 kW, while the wave energy converter (WEC) generated a nominal output of 200 kW. The facility design was flexible enough to integrate a diesel generator with electrochemical storage.

According to Weiner et al. (2001), tests conducted on an RO facility in Israel that was driven by PV and wind energy demonstrated significant results. The planned lifespan was 15 years, and the production rate was calculated to be 3 m^3/day with an expected lifespan. A diesel generator and a battery storage system that can last for two days were built to provide a backup for the system. The investigation found that a 30–50% power ratio between desalination system power and peak power of wind or solar energy was the most efficient.

In 1984, a prototype RO facility powered by wind was developed in the Canary Islands, the foremost European nation for desalination. When wind power was insufficient for plant operation, it was linked to the grid as a supplementary energy source. The output rate was 200 m^3/d, while the energy usage was 5 kWh/m^3 (Article, 2014). In 1993, a hybrid diesel-wind RO facility with a capacity of 56 m^3/day commenced operations at Pájara, Fuerteventura Island. This facility provides the local community with freshwater and electricity. A wind-RO facility without a battery has commenced operations at the same site. Two diesel engines and a WT with 225 kW of power make up the setup (Ma & Lu, 2011). An independent wind-RO system with capacities between 5 and 50 m^3/d was created by the Canary Islands Technological Institute, the AEROGEDESA concept. This approach was established based on substantial experience in the Canary Islands (Article, 2014).

In 2001, as part of the JOULE Program, the University of Las Palmas built and tested a wind/RO system. There was no energy storage in this system. A water/gas (W/G) system typically needs 30 kW of power, but a RO unit would have a 43 to 113 m^3/h capacity (MC Garg, 2019).

In order to supply one thousand people with sufficient amounts of clean water and electricity, Sigarchian et al. (2015) created a small-scale unit of a hybrid energy system that included PV, wind, and engine energy. The primary purpose of this study was to build a transportable container that has the potential to serve as an alternative module for the delivery of electricity and freshwater in the case of a catastrophe. The 22 solar PV modules, windmill, diesel generator, 6 12 V and 200 Ah batteries, and 200 RO units made up the hybrid PV/wind/engine energy-powered RO desalination unit. The modules' rated powers ranged from 5 kW to 1.5 kW. The total power of the unit was 100 kW. TRNSYS was used to develop a model of the system, and its performance was assessed throughout one operating cycle. This evaluation considered two geographical areas that received relatively high solar insolation levels. According to the simulation's findings, solar energy is responsible for more than 63% of the power generation in Nairobi, Kenya, and 80% in Nyala, Sudan. On the other hand, the power generation from wind accounts for 27% and 12% of the total amount of electricity produced in these two cities, respectively. An algorithm for control was implemented to guarantee the dependability and accessibility of the power supply already in place.

A solar-wind hybrid desalination plant was optimized by Mokheimer et al. (2013) after they created a model, ran simulations, and tested several configurations. The planned integrated plant was tested using a representative year's weather data in Saudi Arabia. To minimize average energy expenses, the simulation showed the most efficient way to power a 1 kW RO system running for 12 or 24 hours daily. An average levelized cost of 0.624 USD/kW h was achieved using the optimal arrangement comprising two WTs, 40 solar panels, and six batteries. 66 PV modules, 6 WTs, and 16 batteries were utilized most efficiently to meet a 24-hour power demand of 1 kW. An energy-levelized cost of 0.672 USD/kWh was the lowest. Depending on the raw water's salinity, the desalination process consumes 8–20 kWh/m^3 of energy.

Ghaithan et al. (2021) utilized a multi-objective optimization approach to determine the optimal size for a wind-PV grid-connected system of the highest efficiency. The requirements for energy that an RO unit were considered during the design and development of this system. The overall life cycle cost was the optimization's primary target, and its primary purpose was to minimize it. One of the secondary goals was to lessen the amount of greenhouse gases released by the proposed system. It is essential to point out that the grid was utilized as a supplemental energy source to correct any shortcomings that may have been present in the energy generated from RES. The results demonstrated that a specific energy consumption of 3 kWh/m^3 was associated with the combined system's total capacity of 800 m^3/day.

Ayhan and Madani (2010) suggested the integration of a hybrid wind and solar RE system to generate electricity for the natural vacuum desalination of saltwater in Bahrain. Solar thermal collectors would supply heat to vaporize saltwater at reduced pressure, while a WT may generate power for operating pumps and fans in the system. The projected expense for obtaining freshwater from a system with a daily capacity of 0.13 m^3 is 0.70 USD/m^3 (Ayhan & Al-Madani, 2007).

Ba-Alawi et al. (2023) proposed a novel sizing approach to coordinate the dimensions of RO desalination facilities powered by solar and wind energy. Furthermore, they employed a follower-leader optimization framework to enhance the overall design, with or without the integration of demand-side water management (DSWM). WTs and PV panels generated electricity, a battery bank stored surplus energy, an RO system desalinated seawater to produce clean water, and storage tanks held the freshwater.

A hydrogen storage system was implemented to retain surplus electricity produced at the conclusion of each day for long-term storage. The RO system primarily derives its energy from photovoltaic (PV) systems and WTs, both of which are RES. A battery-hydrogen storage (BHS) unit was employed to retain surplus electricity produced by the system.

Khalilzadeh and Nezhad (2018) conducted a sensitivity study to assess the impact of several characteristics, including wind speed, on overall efficiency. The high-capacity windmill produced significant waste heat, varying from 122 kW at a wind velocity of 8 meters per second to 269.9 kW at 12 meters per second. The multi-effect distillation (MED) unit can generate freshwater at a rate of 23.81 m^3/day by employing a wind speed of 8 meters per second. This is accomplished by harnessing the waste heat that is generated by the WT. Increasing the wind speed was also shown to have the potential to improve the combined system's energy efficiency, as was previously mentioned. At a speed of 12 meters per second, the efficiency increased from 69.14% at 8 meters per second to 46.52%.

Additionally, by increasing the wind speed, it is possible to improve the exergy efficiency of the combined system for the combined system. Specifically, it can increase from 7.34% at 8 meters per second to 13.30% at 12 meters per second. It has been determined through the thermo-economic analysis that the integrated system will result in an anticipated cost of 16.676 USD/m^3 for the production of drinkable water.

5.6 CONCLUSION

The chapter highlights the critical interconnection between energy and water resources, emphasizing the growing need for sustainable solutions to address global water scarcity. It underscores the advancements in desalination technologies and the pivotal role of RE, particularly wind power, in making desalination more viable and environmentally friendly. The chapter details various methods of integrating wind energy with desalination processes, showcasing successful case studies and technological innovations. It also discusses the economic and feasibility aspects, stressing the importance of optimizing system components and configurations to reduce costs and enhance efficiency. The chapter adopts wind-powered desalination systems as a sustainable and cost-effective solution to meet the increasing demand for freshwater in regions with high wind potential.

REFERENCES

Article, C. (2014). *Case Article Desalination: Case study of Lakshadweep Islands and Gran Canaria Islands*. *4*(2), 1073–1078.

Ayhan, T., & Al-Madani, H. (2007). *Analysis of an innovative solar water desalination system using gravity induced vacuum.* https://www.osti.gov/etdeweb/biblio/20968165

Ayhan, T., & Madani, H. (2010). Feasibility study of renewable energy powered seawater desalination technology using natural vacuum technique. *Renewable Energy.* https://www.sciencedirect.com/science/article/pii/S0960148109002833

Ba-Alawi, A., Nguyen, H., & Yoo, C. (2023). Sustainable design of a solar/wind-powered reverse osmosis system with cooperative demand-side water management: A coordinated sizing approach with a fuzzy. *Energy Conversion and Management.* https://www.sciencedirect.com/science/article/pii/S0196890423009706

Bundschuh, J. & Hoinkis, J. (2019). *Renewable Energy Applications for Freshwater.* CRC Press.

Bundschuh, J., Kaczmarczyk, M., Ghaffour, N., & Tomaszewska, B. (2021). State-of-the-art of renewable energy sources used in water desalination: Present and future prospects. *Desalination, 508*(March). https://doi.org/10.1016/j.desal.2021.115035

Cabrera, P., Folley, M., & Carta, J. A. (2021). Design and performance simulation comparison of a wave energy-powered and wind-powered modular desalination system. *Desalination, 514*(May). https://doi.org/10.1016/j.desal.2021.115173

Dahioui, Y., & Loudiyi, K. (2013). Wind powered water desalination. *Proceedings of 2013 International Renewable and Sustainable Energy Conference, IRSEC 2013*, 257–262. https://doi.org/10.1109IRSEC.2013.6529659

Damania, R., Desbureaux, S., Hyland, M., Islam, A., Moore, S., Rodella, A.-S., Russ, J., & Zaveri, E. (2017). Uncharted Waters: The New Economics of Water Scarcity and Variability. In *Uncharted Waters: The New Economics of Water Scarcity and Variability.* https://doi.org/10.1596/978-1-4648-1179-1

Darwish, A. S. K. (2021). Wind energy utilisation for water desalination, street and buildings lighting – a case study for The Emirate of Ajman – UAE. *Renewable Energy and Environmental Sustainability, 6*, 10. https://doi.org/10.1051/rees/2021012

Elimelech, M., & Phillip, W. A. (2011). The future of seawater desalination: Energy, technology, and the environment. *Science, 333*(6043), 712–717. https://doi.org/10.1126/science.1200488

Eltawil, M. A., Zhengming, Z., & Yuan, L. (2008). Wind energy technologies integrated with desalination systems: Review and state-of-the-art. *12*, 1099. https://www.sciencedirect.com/science/article/pii/S0011916411003651

Eltawil, M. A., Zhengming, Z., & Yuan, L. (2009). A review of renewable energy technologies integrated with desalination systems. *Renewable and Sustainable Energy Reviews, 13*(9), 2245–2262. https://doi.org/10.1016/j.rser.2009.06.011

Forstmeier, M., Mannerheim, F., D'Amato, F., Shah, M., Liu, Y., Baldea, M., & Stella, A. (2007). Feasibility study on wind-powered desalination. *Desalination, 203*(1–3), 463–470. https://doi.org/10.1016/J.DESAL.2006.05.009

Francisco, F., Leijon, J., Boström, C., Engström, J., & Sundberg, J. (2018). Wave power as solution for off-grid water desalination systems: Resource characterization for Kilifi-Kenya. *Energies, 11*(4), 1004. https://doi.org/10.3390/en11041004

Fytikas, M., Radoglou, G., Karytsas, C., Mendrinos, D., Vasalakis, A., & Andritsos, N. (2005). Geothermal research in Vounalia Area, Milos Island, for seawater desalination and power production. *Proceedings World Geothermal Congress, 2*, 24–29.

García-Rodríguez, L. (2003). Renewable energy applications in desalination: State of the art. *Solar Energy, 75*(5), 381–393. https://doi.org/10.1016/j.solener.2003.08.005

García-Rodríguez, L. (2004). Desalination by wind power. *Wind Engineering, 28*(4), 453–463. https://doi.org/10.1260/0309524042886405

Garg, M. C. (2019). Renewable energy-powered membrane technology: cost analysis and energy consumption. *Current Trends and Future Developments on (Bio-) Membranes.* https://www.sciencedirect.com/science/article/pii/B9780128135457000040

GE Global Research. (2004). *Integrated Wind Energy/Desalination System: October 11, 2004 - July 29, 2005. October.* http://www.osti.gov/bridge

Ghaffour, N., Bundschuh, J., Mahmoudi, H., & Goosen, M. F. A. (2015). Renewable energy-driven desalination technologies: A comprehensive review on challenges and potential applications of integrated systems. *Desalination, 356*, 94–114. https://doi.org/10.1016/j.desal.2014.10.024

Ghaffour, N., Lattemann, S., Missimer, T., Choon Ng, K., & Amy, G. (2013). Seawater desalination using renewable energy: solar, geothermal, and wind 1 2. In *Proceeding of International Conference on Applied Energy (ICAE 2013).*

Ghaffour, N., & Mujtaba, I. M. (2018). Desalination using renewable energy. In *Desalination* (Vol. *435*, Issue 96). Center for Renewable Energy Sources. https://doi.org/10.1016/j.desal.2018.01.029

Ghaithan, A., Al-Hanbali, A., Mohammed, A., Attia, A. et al. (2021). *Optimization of a solar-wind-grid powered desalination system in Saudi Arabia.* Elsevier. https://www.sciencedirect.com/science/article/pii/S0960148121009265

Giwa, A. A. A. (2015). Wind and Geothermal Energy in Desalination: A Short Review on Progress and Sustainable Commercial Processes. *Industrial Engineering & Management, 4*(4), 175. https://doi.org/10.4172/2169-0316.1000175

Goosen, M., Mahmoudi, H., & Ghaffour, N. (2024). Today's and future challenges in applications of renewable energy technologies for desalination. *Critical Reviews in Environmental Science and Technology, 44*(9), 929–999. https://doi.org/10.1080/10643389.2012.741313

Greco, F., Heijman, S. G. J., & Jarquin-Laguna, A. (2021). Integration of wind energy and desalination systems: A review study. *Processes, 9*(12), 1–29. https://doi.org/10.3390/pr9122181

Heidary, B., Hashjin, T. T., Ghobadian, B., & Roshandel, R. (2018). Optimal integration of small scale hybrid solar wind RO-MSF desalination system. *Renewable Energy Focus, 27*(00), 120–134. https://doi.org/10.1016/j.ref.2018.05.003

Hossam-Eldin, A. A., Youssef, K. H., & Kotb, H. (2017). Technical and economic optimization of reverse osmosis desalination systems integrated with PV/wind energy resources: A case study. *2017 19th International Middle-East Power Systems Conference, MEPCON 2017 - Proceedings, 2018-Febru* (March 2018), 510–514. https://doi.org/10.1109/MEPCON.2017.8301228

Jarquin-Laguna, A., & Greco, F. (2019). Integration of hydraulic wind turbines for seawater reverse osmosis desalination. *2019 Offshore Energy and Storage Summit, OSES 2019*, 1–9. https://doi.org/10.1109/OSES.2019.8867343

Karameldin, A., Lotfy, A., & Mekhemar, S. (2003). The Red Sea area wind-driven mechanical vapor compression desalination system. *Desalination.* https://www.sciencedirect.com/science/article/pii/S0011916402010925

Karytsas, C., Mendrinos, D., & Radoglou, G. (2004). The current geothermal exploration and development of the geothermal field of Milos island in Greece. *GHC Bulletin, 25*(June), 17–21.

Kershman, S., Rheinländer, J., & Gabler, H. (2003). Seawater reverse osmosis powered from renewable energy sources-hybrid wind/photovoltaic/grid power supply for small-scale desalination in Libya. *Desalination.* https://www.sciencedirect.com/science/article/pii/S0011916402010895

Khalilzadeh, S., & Nezhad, A. (2018). Utilization of waste heat of a high-capacity wind turbine in multi effect distillation desalination: Energy, exergy and thermoeconomic analysis. *Desalination.* https://www.sciencedirect.com/science/article/pii/S001191641732307X

Koklas, P., & Papathanassiou, S. (2006). Component sizing for an autonomous wind-driven desalination plant. *Renewable Energy.* https://www.sciencedirect.com/science/article/pii/S0960148105002855

Koroneos, C., Dompros, A., & Roumbas, G. (2007). Renewable energy driven desalination systems modelling. *Journal of Cleaner Production, 15*(5), 449–464. https://doi.org/10.1016/j.jclepro.2005.07.017

Koutroulis, E., & Kolokotsa, D. (2010). Design optimization of desalination systems power-supplied by PV and W/G energy sources. *Desalination*. https://doi.org/10.1016/j.desal.2010.03.018

Kumar, L., Soomro, J., Khoharo, H., & El Haj Assad, M. (2023). A comprehensive review of solar thermal desalination technologies for freshwater production. *AIMS Energy, 11*(2), 293–318. https://doi.org/10.3934/ENERGY.2023016

Lai, W., Ma, Q., Lu, H., Weng, S., Fan, J., & Fang, H. (2016). Effects of wind intermittence and fluctuation on reverse osmosis desalination process and solution strategies. In *Desalination* (Vol. *395*, pp. 17–27). https://doi.org/10.1016/j.desal.2016.05.019

Liu, C. C. K., Jae-Woo, P., Migita, R., & Gang, Q. (2002). Experiments of a prototype wind-driven reverse osmosis desalination system with feedback control. *Desalination, 150*(3), 277–287. https://doi.org/10.1016/S0011-9164(02)00984-0

Lotfy, H. R., Staš, J., & Roubík, H. (2022). Renewable energy powered membrane desalination — review of recent development. *Environmental Science and Pollution Research, 29*(31), 46552–46568. https://doi.org/10.1007/s11356-022-20480-y

Ma, Q., & Lu, H. (2011). Wind energy technologies integrated with desalination systems: Review and state-of-the-art. *Desalination, 277*(1–3), 274–280. https://doi.org/10.1016/j.desal.2011.04.041

Maftouh, A., Fatni, O. El, Bouzekri, S., Bahaj, T., Kacimi, I., Hajjaji, S. El, & Malik, A. (2023). Solar desalination: Current applications and future potential in MENA region – A case study. *Journal of Sustainable Development of Energy, Water and Environment Systems, 11*(2), 1–26. https://doi.org/10.13044/j.sdewes.d10.0437

Manju, S., & Sagar, N. (2017). Renewable energy integrated desalination: A sustainable solution to overcome future fresh-water scarcity in India. *Renewable and Sustainable Energy Reviews, 73*, 594–609. https://doi.org/10.1016/j.rser.2017.01.164

Mbarga, A. A., Song, L., Ross Williams, W., & Rainwater, K. (2014). Integration of Renewable Energy Technologies With Desalination. *Current Sustainable/Renewable Energy Reports, 1*(1), 11–18. https://doi.org/10.1007/s40518-013-0002-1

Mersin, G., & Çeliktaş, M. S. (2023). Integration of Renewable Energy Systems. In *Handbook of Smart Energy Systems* (pp. 2401–2424). https://doi.org/10.1007/978-3-030-97940-9_93

Mokheimer, E., Sahin, A., Al-Sharafi, A., & Ali, A. (2013). Modeling and optimization of hybrid wind–solar-powered reverse osmosis water desalination system in Saudi Arabia. *Energy Conversion and Management*. https://www.sciencedirect.com/science/article/pii/S0196890413002987

Novosel, T., Ćosić, B., Pukšec, T., Krajačić, G., Duić, N., Mathiesen, B. V., Lund, H., & Mustafa, M. (2014). Integration of renewables and reverse osmosis desalination - Case study for the Jordanian energy system with a high share of wind and photovoltaics. *Energy, 92*(Part 3), 270–278. https://doi.org/10.1016/j.energy.2015.06.057

Oré, D. G., Pretell, V. H., & Pilco, A. W. (2022). Design of an autonomous photovoltaic system for desalination of groundwater by electrodialysis. *Journal of Physics: Conference Series, 2180*(1). https://doi.org/10.1088/1742-6596/2180/1/012020

Petrovic, A., Delibasic, B., Filipovic, J., Petrovic, A., & Lomovic, M. (2018). Thermoeconomic and environmental optimization of geothermal water desalination plant with ejector refrigeration system. *Energy Conversion and Management, 178*(October), 65–77. https://doi.org/10.1016/j.enconman.2018.10.035

Rashidi, M. M., Mahariq, I., Murshid, N., Wongwises, S., Mahian, O., & Alhuyi Nazari, M. (2022). Applying wind energy as a clean source for reverse osmosis desalination: A comprehensive review. *Alexandria Engineering Journal*, *61*(12), 12977–12989. https://doi.org/10.1016/j.aej.2022.06.056

Rodríguez, G., Rodríguez, M., Pérez, J., & Veza, J. (1996). A systematic approach to desalination powered by solar, wind and geothermal energy sources. In *Proceedings of the Mediterranean Conference on Renewable Energy Sources for Water Production. European Commission, EURORED Network, CRES, EDS* (pp. 20–25). https://scholar.google.com/scholar?lookup=0&q=Rodriguez-Girones,+Rodriguez+M,+Perez+J,+Veza+J.+A+systematic+approach+to+desalination+powered+by+solar,+wind+and+geothermal+energy+sources&hl=en&as_sdt=0,5#0

Romero-Ternero, V., García-Rodríguez, L., & Gómez-Camacho, C. (2005). Thermoeconomic analysis of wind powered seawater reverse osmosis desalination in the Canary Islands. *Desalination*. https://www.sciencedirect.com/science/article/pii/S0011916405006995

Rybar, S., Vodnar, M., Vartolomei, L., León Méndez, R., Blas, J., & Ruano, L. (2005). Experience with renewable energy source and SWRO desalination in Gran Canaria. *Fgwater.Com*, 5–100. https://www.fgwater.com/static/upload/file/20230615/1686811256486630.pdf

Saravana, B. S., & Rajini, V. (2014). Wind Energy Integration with Desalination Industry – Analysis of Wind Power Potential in Chennai, Nagapattinam and Puducherry of South India. 187–190. https://inpressco.com/special_edition/wind-energy-integration-with-desalination-industry-analysis-of-wind-power-potential-in-chennainagapattinam-and-puducherry-of-south-india/

Serrano-Tovar, T., Peñate Suárez, B., Musicki, A., de la Fuente Bencomo, J. A., Cabello, V., & Giampietro, M. (2019). Structuring an integrated water-energy-food nexus assessment of a local wind energy desalination system for irrigation. *Science of the Total Environment*, *689*, 945–957. https://doi.org/10.1016/j.scitotenv.2019.06.422

Sigarchian, S., Paleta, R., Malmquist, A., & Pina, A. (2015). Feasibility study of using a biogas engine as backup in a decentralized hybrid (PV/wind/battery) power generation system–Case study Kenya. *Energy*. https://www.sciencedirect.com/science/article/pii/S0360544215009020

Tzen, E., & Morris, R. (2003). Renewable energy sources for desalination. *Solar Energy*, *75*(5), 375–379. https://doi.org/10.1016/j.solener.2003.07.010

Weiner, D., Fisher, D., Moses, E., Katz, B., & Meron, G. (2001). Operation experience of a solar-and wind-powered desalination demonstration plant. *Desalination*. https://www.sciencedirect.com/science/article/pii/S0011916401001989

Zheng, S., & Qian, G. (2019). Overview of hydro-wind-solar power complementation development in China. *Global Energy Interconnection*, *2*(4), 285–289. https://doi.org/10.1016/j.gloei.2019.11.011

Zuo, L., Chen, J., Zhou, X., & Xu, B. (2019). A preliminary study of wind supercharged solar chimney power plant combined with seawater desalination. *Taiyangneng Xuebao/Acta Energiae Solaris Sinica*, *40*(3), 620–627. https://www.sciencedirect.com/science/article/pii/S0038092X1830272X

Photovoltaic RO and Electrodialysis 6

6.1 INTRODUCTION

Water serves as an essential constituent for sustaining life on the planet Earth. Current conditions in the water environment are characterized by significant contamination, resulting in substantial degradation of natural water courses (Al-Karaghouli et al., 2010). Consequently, a significant decline is observed in expanding the natural environment, human well-being, and economic prosperity (Petersen et al., 2018). In order to protect natural water sources from contamination, rules about domestic water and wastewater usage in urban areas have been implemented. Ongoing health issues and environmental challenges are caused by poor handling of resident wastewater. This is because organic molecules are not adequately removed, including toxic and non-toxic chemicals, viruses, and pathogenic microbes (Al-Amshawee et al., 2020). Industrial wastewater comprises various organic and inorganic chemicals, including oil, insecticides, pollutants, hydrocarbons, aromatic, and dyes (Andersson, 2009). These constituents play a crucial role in the occurrence of water pollution. Consequently, wastewater companies strive to maintain water resources by implementing a mediation process characterized by a high organic loading rate, low cost, simplicity, feasibility, and safety (Al-Amshawee et al., 2020).

Water and energy are two precious resources. The scarcity and high cost of freshwater in arid locations contribute significantly to the problem of insufficient access to potable water (Panagopoulos, 2021). Conventional desalination systems, either membrane or thermal, need substantial energy input. Using fossil fuels for desalination is deemed unsustainable and economically impractical in the long run due to increasing expenses and scarcity of these fuels (Ettouney & Wilf, 2009). Furthermore, their usage contributes to the emission of greenhouse gases and the dispersion of other pollutants and particulate matter. Integrating renewable energy sources (RES) into the desalination business is crucial for current and future environmental conservation efforts (Jeff, 2019).

 DOI: 10.1201/9781003530893-6

In most cases, desalination units may be included in polygeneration facilities, integrated systems producing several energy forms (Luqman et al., 2020). Waste heat from thermal power plants has been effectively used to drive thermal desalination facilities or to warm salt water inputs. Incorporating desalination systems into polygeneration systems is highly appealing, particularly when these systems are combined with RES. Various options are available for incorporating desalination technology into renewable polygeneration systems. Thermal energy and electrical energy are potential power sources for desalination plants (Al-Sahali & Ettouney, 2007; El Haj Assad et al., 2021).

Thermal energy-driven desalination technologies are well-suited for installation in large-scale district-level polygeneration systems (Belessiotis et al., 2016). On the other hand, desalination units powered by electrical energy are much more adaptable and may be used for both large-scale and small systems. The appeal of large-scale desalination systems depends on their potential to provide enhanced economic profitability and increased operational efficiency (Reif & Alhalabi, 2015). Evaluating the system risks of large-scale projects is essential to assess their viability accurately. Polygeneration systems utilizing thermally based desalination methods can demonstrate significant complexity, requiring sophisticated control and management technology to enhance system performance efficiently (Mannan & Al-Ghamdi, 2019). Thermally driven desalination systems are frequently used with renewable energy, such as geothermal and solar energy. Hence, these systems often exhibit substantial capacity and are very appealing for generating energy (thermal, electrical, and cooling) and replenishing freshwater at the district level (Kalidasan et al., 2021). Polygeneration systems that rely on electric renewable sources, such as photovoltaic (PV) or wind turbines, present a promising situation within this context (Kumar et al., 2023). The main aim of this chapter is to find the possibility of combining PV systems with electrodialysis (ED) and reverse osmosis (RO) desalination processes. It demonstrates the potential of solar power to deliver drinkable water in desert locations, highlighting the benefits of integrating renewable energy with desalination systems to solve the issue of reduced dependency on fossil fuels and the issue of water scarcity.

6.2 PHOTOVOLTAIC

PV conversion involves converting solar radiation directly into electrical current without needing a heat engine (Karavas et al., 2019). One of the most significant advantages of PV devices is that they can be constructed as individual units, which allows them to provide outputs that range from microwatts of energy to megawatts of energy. These devices are also simple in design and require minimal maintenance (Oré et al., 2022). For this reason, they are used as a power source in remote structures, to pump water in solar household systems, spacecraft and satellites, RO plants, and even megawatt-scale power stations. Demand for PVs continues to rise annually due to its wide range of applications (Garg & Joshi, 2015).

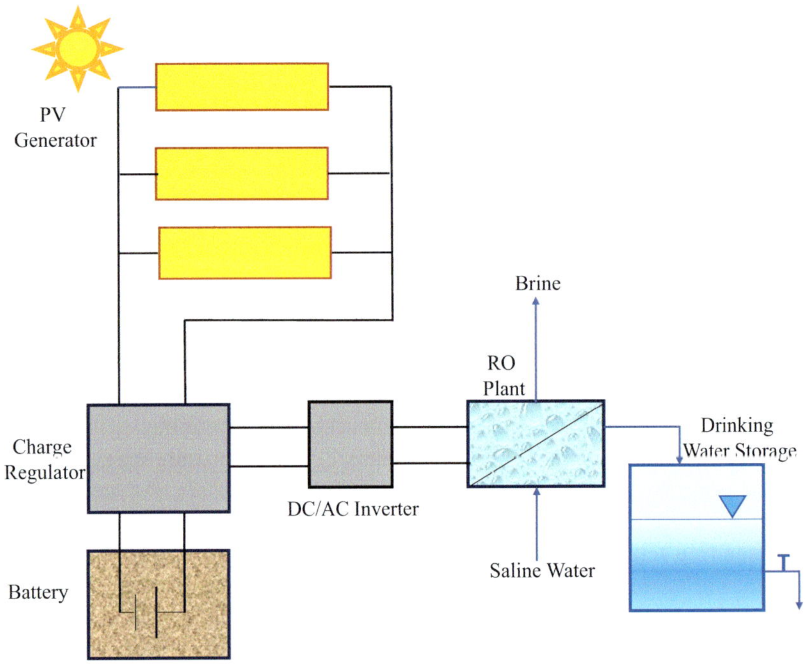

FIGURE 6.1 Photovoltaic RO desalination system.

Figure 6.1 Diagram showing a PV RO desalination system. Solar panels are connected with a charge regulator and battery, connected to a DC/AC inverter. Saline water enters the RO plant, separating into brine and drinking water stored in a tank.

Solar thermal energy is a widely utilized technology currently being investigated as a potential alternative (AlMadani, 2003). Usually, thermal and concentrated collectors are utilized to produce the energy required. The radiant thermal energy emitted by the sun is harnessed through various means, such as concentrated solar power (CSP), which involves extracting thermal energy from the sun (A. H. Galama et al., 2014; Karavas et al., 2019). Another method is concentrator photovoltaic (CPV)/photovoltaic (PV), which utilizes solar modules to convert photon-carried solar energy into electrical energy (Ortiz et al., 2008). This chapter examines the diverse investigations related to the exploration of desalination techniques utilizing solar collectors to desalinate seawater and the synergistic cooling, heating, and electrical processes based on PVs (Karimi et al., 2015).

6.2.1 Photovoltaic RO

Solar PV energy can power an RO membrane desalination system to produce potable water (Herrero-Gonzalez et al., 2024). This is especially useful in outlying locations with a severe lack of water, high salt, and no access to the electric grid (Elfaqih et al., 2024) (Figure 6.1).

Thomson and Infield proposed a PV RO plant that operates without a battery. The model was verified, and the system's capacity of 3 m^3/day, utilizing a 2.4 kWp PV array, was regulated through tests in the lab (Bermudez-Contreras et al., 2008). A small RO (RO) system that runs on solar (PV) energy was developed and constructed by Joyce et al. (2001) for usage in remote rural regions. Operating at pressures as low as 5 bar, this plant could daily produce 100–500 L of RO water. Using a simple spreadsheet model, Qasim et al. (2019) showed that hybrid wind-PV technology with energy recovery capabilities may power a seawater RO desalination plant.

An energy recovery device of the pressure-exchanger type achieved approximately 50% savings in actual energy use. In their study, Richards and Schäfer (2003) used a solar PV-powered NF-UF-RO system to analyze the effects of different energy levels and pH on removing dissolved contaminants from real groundwater in Australia. The sustainable power membrane system consistently eliminated salts and inorganic pollutants under diverse natural energy and pH settings through diffusion, convection, and precipitation processes. This has major consequences for water use in isolated areas, where pollutants that are resistant to conventional removal methods pose a threat. A small-sized autonomous solar-PV RO system was simulated by Clarke et al. (2013) to investigate the benefits and drawbacks of using and not using battery storage. Power characteristics of RO systems and the time-based accuracy of simulations were shown to have varying effects on system efficiency. A compact desalination model (1 m^3/d) was developed and tested by Soric et al. (2012). It used rooftop solar panels for powering an RO process; however, it did not include an intermediate storage battery.

There are several benefits to combining PV with desalination operations. It offers a long-term solution for areas with water scarcity while simultaneously decreasing emissions of greenhouse gases and dependence on fossil fuels. Solar PV desalination relies on energy efficiency. One method that has shown promise in reducing power usage is energy recovery from RO systems' concentrate stream. The usefulness and efficiency of PV desalination have been proven in case studies. For instance, in rural locations, communities have relied on small-scale RO/PV systems to deliver water reliably (Darwish et al., 2016). Typical components of such systems driven by PV arrays include pretreatment units, RO vessels, high-pressure pumps, and storage tanks. Using advanced cell technology units (ACCUs) ensures that the system will continue to run even when solar irradiation is low.

PV systems that utilize RO to desalinate water have been the subject of several studies (Ortiz et al., 2007). An RO seawater desalination plant powered by PVs was built in Durban, South Africa, by Ndadane and Inambao (2022). Increasing the amount of freshwater created annually, the proposed plant can use three different power sources solar energy, concentrator and battery, and wind to generate 16,830,000 L of water. The 6 kW of PV power it generated further demonstrated the RO's portability. By powering it with solar energy, RO desalination becomes a greener and more economical option.

Freire-Gormaly and Bilton (2019a) conducted a performance assessment of the membrane regarding scaling and fouling over the operation of the PV-RO device under three distinct conditions: the use or non-use of anti-scale chemicals, a constant or variable flow rate, and the washing of the membrane with permeate water or the absence of washing. The findings indicated that the frequency of membrane scaling escalates with intermittent processes. Utilizing rinsing maintenance before system closure preserves permeate efficiency. Freire-Gormaly established a statistical model and design system for membrane fouling and a site-specific PV-RO system. A laboratory-scale system was employed to examine continuous and intermittent procedures and the anti-scalant and rinsing processes (Freire-Gormaly and Bilton, 2019b).

Karavas et al. (2019) created a PV-powered small-scale seawater reverse osmosis (SWRO) desalination machine that is now being studied in the Cyclades complex to find the best configuration for self-sufficient operation from a technical and economic standpoint. The best system was defined after five potential case study situations were considered. No energy storage system was used in the first case, while a nominal load was used in the second. An energy management method was needed to ascertain the desalination unit's operating point regarding the PV power that was accessible in the third situation, which involved a battery bank. The fifth and fourth cases used fuzzy logic, fuzzy cognitive maps, and computational intelligence approaches. Case five, which showed the best system in terms of cost and power losses relative to commonly employed tactics in commercial RO plants, was the best option. Over a 20-year investment horizon, the technology also demonstrated a reduced cost. Using a standard energy storage system with a limited capacity and an EMS that utilizes fuzzy logic, the suggested PV-SWRO desalination plant may optimize potential solar panel energy and provide acceptable quality freshwater under all operating situations. This technology provides an eco-friendly, practical, cost-effective alternative to island water scarcity.

Karimi et al. (2015) examined the viability of utilizing renewable energy to operate desalination plants in remote, rural areas. They performed pilot-scale tests and software simulations to ascertain the energy usage of electrodialysis reversal (EDR) and RO systems across varying water salinity levels, flow rates, and temperatures. The findings indicated that the product flow rate and feed water saltiness substantially influenced the specific energy utilization of EDR and RO. The research indicated that temperature fluctuations did not markedly impact the specific energy consumption (SEC) of EDR but considerably affected the SEC of RO. The researchers employed HOMER software to determine the most effective renewable energy solutions for powering desalination equipment. The research indicated that solar energy might effectively energize off-grid EDR-ED and RO plants in areas with comparable atmospheric conditions.

Voutetaki et al. (2022) developed and examined a photovoltaic-electrodialysis system for the remediation of nitrate-contaminated water in Greece. The concentration of nitrate in drinking water dropped to less than 50.0 mg/L, as it was roughly 100 mg/L. Based on the data, it was found that 130 kW were needed to purify 4 m^3 of water with 100 mg/L of nitrate after 68 pairs of ED cells running at 25 V were installed. The energy was generated using various photovoltaic systems powered by batteries. The two alternatives considered were the most suitable: ten sun panels and five batteries, producing the most water. The second choice, which produced the least surplus energy, involved nine solar panels and six batteries. The research determined that PV-ED systems benefit the environment and can clean nitrate-contaminated water, which is particularly important for remote locations without access to the power grid.

Meratizaman and Godarzi's (2019) work evaluates the financial feasibility of Iran's solar-powered RO desalination facility, utilizing lithium battery energy storage. The facility seeks to mitigate freshwater scarcity in arid areas, especially in the Middle East, where harsh climates and increasing populations necessitate desalination. The research employs solar radiation models, lithium batteries, PV solar panels, and RO desalination devices to assess the economic feasibility of the system. The principal system comprises a 1.4 m^3/day brackish water (BW) RO plant using two membrane modules and four PV panels. The energy storage system employs 24 V5 200 Ah lithium batteries. The research determined that the approach is economically feasible, yielding the most profits within five years. The research emphasized the significance of local climate and economic conditions in establishing and operating renewable energy-powered desalination facilities.

Bian and Jun (2017) have enhanced a PV-integrated ED reversal device for small-scale use in rural India. The technology necessitates 50% of the specific energy and diminishes water wastage from 60% to under 10% compared to smaller-scale RO units. The ideal system costs $23,420, which is 42% lower than traditional engineering practices. The essential factor in cost reduction was adaptable water production that responds to daily variations in solar irradiation and incorporates water buffer storage tanks. A sensitivity study indicated that the capital cost of the entire system is predominantly influenced by membrane area, with an 87% reduction in membrane cost resulting in a 50% decrease in system capital cost. The optimization technique and cost-reduction measures are relevant to other PV-EDR frameworks and broader off-grid desalination uses.

For farming purposes in the MENA region, Jacob et al. (2022) provide PV-integrated ED reversal, a promising desalination system. Due to water scarcity, BW irrigates crops throughout the Middle East and North Africa. This practice has several drawbacks, such as lower agricultural yields and soil damage. There is an over-reliance on fossil fuels to power our homes and businesses, and renewable water resources are being depleted due to the current solutions' high water and energy demands. Energy cost is the biggest problem with a smaller-scale desalination unit in both regions. When compared to on-grid RO processes, PV-EDR uses less energy, making it a better desalination design. We introduce a new time-variable operational method for continuous PV-integrated EDR, enabling low-energy, battery-free, continuous solar desalination. Despite a higher initial investment, a case study in Jordan found that PV-integrated EDR had a levelized cost of water (LCOW) 24% lower than existing RO systems. More efficient and effective desalination with better recovery could be possible using TV-PV-EDR in the agricultural sector.

Riffel et al. (2009) built a compact PV RO facility to desalinate BW in equatorial regions. Two operational approaches were simulated and evaluated. The first plant employs two modules for RO pumping and one for well pumping, while the second plant utilizes three modules for both operations. At Station 2, a DC-DC buck converter was constructed to incorporate maximum power level control. The outcomes indicate that Plant 2 exhibits greater efficiency, including a 60% rise in daily permeate generation and a 32% enhancement in daily operational duration, with an average specific

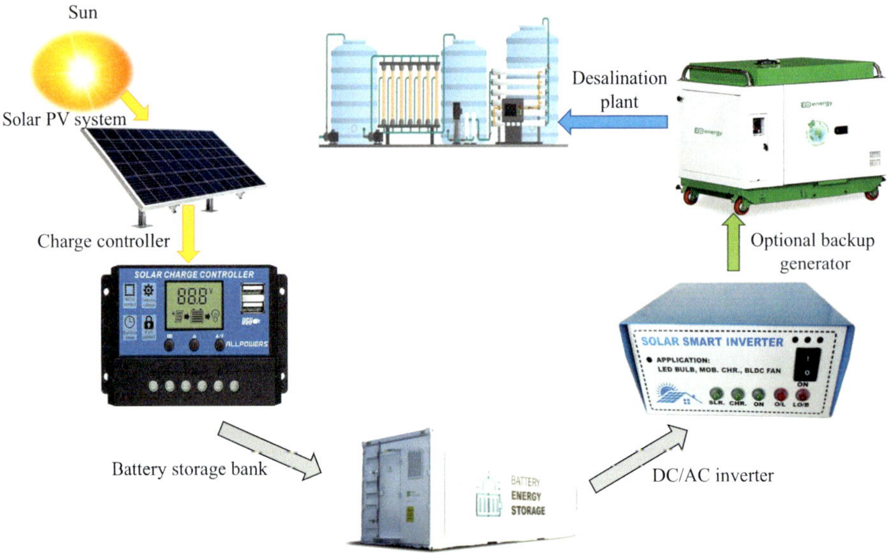

FIGURE 6.2 Photovoltaic unit (off-grid) with battery and desalination system.

energy use of 1.57 kWh.m^3 and a PV power-permeate everyday rate of production of 0.63 Wp.LÀ1 for supply water containing 800.0 mg.LÀ1 of total dissolved solids (TDS) (D. Riffel et al., 2021).

Bayod-Rújula and Martínez-Gracia (2009) introduce an innovative design for simultaneous desalination and electrical generation. The system comprises solar modules arranged in parallel and series, with an ED reactor connected directly to several modules. The PV arrays may be linked to batteries or the electrical grid. The ED system is a time-variable resistance that extracts a time-dependent current from the solar modules. The excess energy produced by the solar panel is integrated into the power system without altering the arrangement. Simulations using PSpice indicate that the suggested system exhibits superior global efficiency compared to traditional designs. A novel technique for regulating the inverter or regulator is proposed, emphasizing the optimal power point (OPP). This technology is appropriate for local energy and water provision in regions with elevated irradiation levels and demand for BW desalination (Figures 6.2 and 6.3).

Figure 6.2 The diagram illustrates an off-grid solar energy system for a desalination plant, featuring a PV unit, charge controller, and battery storage bank. It includes an optional backup generator and DC/AC inverter. Arrows indicate energy flow between components in the desalination system.

Figure 6.3 The illustration depicts a PV system (on-grid) powering a desalination plant. It features solar panels connected to a DC/AC inverter, then to a power meter, seamlessly integrating with the electric power grid. Also, it includes the solar panel, inverter, power meter, grid, and desalination plant.

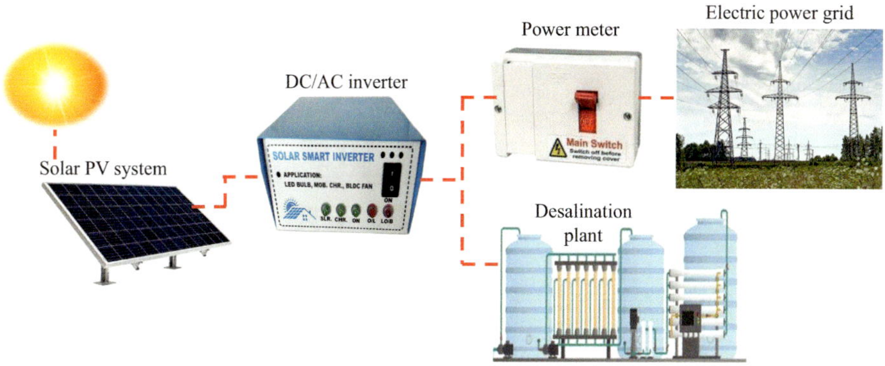

FIGURE 6.3 Photovoltaic system (on-grid) operated desalination plant.

6.3 ELECTRODIALYSIS

ED was commercialized in the early 1960s, approximately a decade before RO (AWWA Manual, 1999). It enabled inexpensive and efficient BW desalination, generating considerable interest in desalination technology for municipal drinking water production (Y. D. Ahdab et al., 2021). Shatat and Riffat (2014) assert that it possesses numerous more applications in biochemistry and environmental conservation besides its use in producing edible salt. Electrochemical separation, commonly referred to as ED, is a technique that employs electrically charged ion-exchange membranes, which possess a difference in electrical potential that serves as the driving force. This is contingent upon the observation that most salts dissolved in water are ionic, signifying that they possess either a negative (anionic) or positive (cationic) charge and consequently migrate toward electrodes with an opposing electric charge (Herrero-Gonzalez et al., 2018). Designing membranes to allow for the selective transit of cations or anions is possible (Figure 6.4).

Figure 6.4 Diagram of an electrochemical process shows saline feedwater entering the system, passing between cation- and anion-selective membranes. Ions move toward electrodes, resulting in concentrate and freshwater outputs. Arrows indicate water and ion flow directions.

Applying an electric field between the electrodes induces the migration of anions and cations toward both negative and positive electrodes due to the resultant ionic current (Valero et al., 2011). In conjunction with the process of selectively separating cations and anions via an arrangement of cation-exchange membranes (CEMs) as well as anion-exchange membranes (AEMs), this ionic transport mechanism leads to the formation of channels, which allow the movement of solutions with varying concentrations, namely concentrate and dilute (AlMadani, 2003). The separation process through ED occurs when cations moving inside the CEM are prevented from reaching the adjacent AEM (Lotfy et al., 2022). As a consequence, the elimination of salt facilitates its subsequent development within concentration paths (Al-Amshawee et al., 2020). An AEM

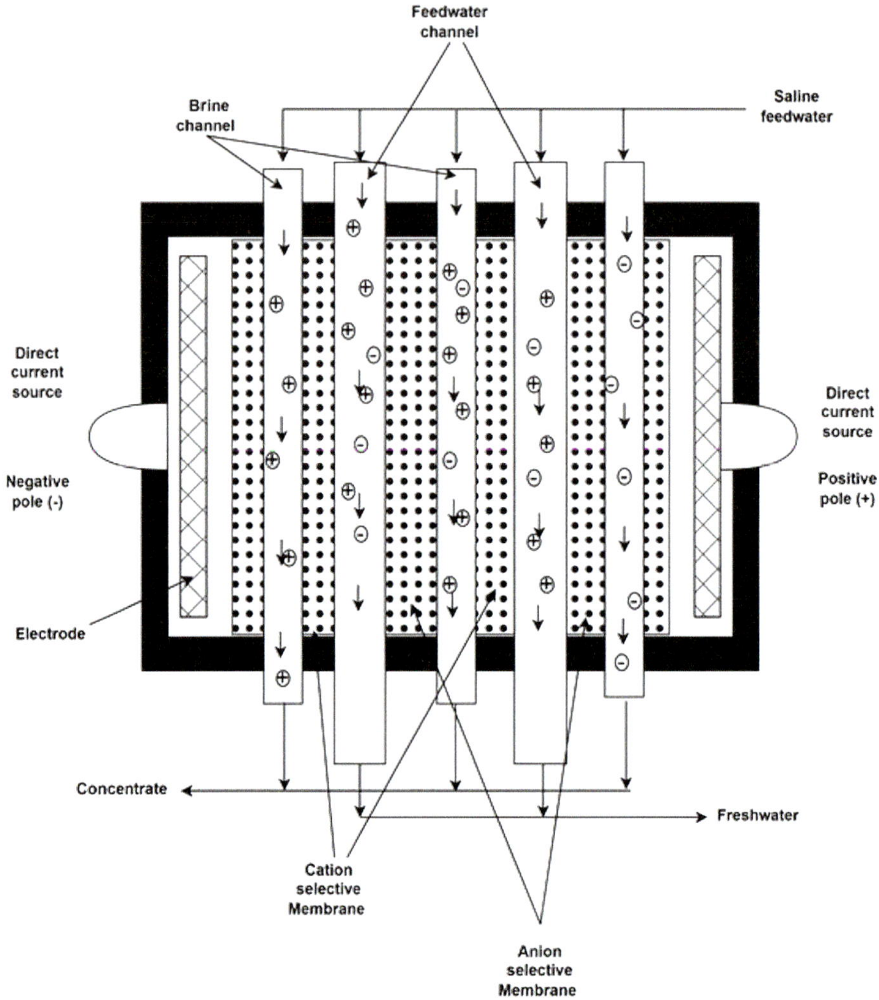

FIGURE 6.4 Electrochemical.

and a CEM, along with their adjacent channels, comprise a single cell pair: the composite. The ion-exchange membranes used in ED have two notable features: reduced resistance to electricity and increased selectivity (Kuroda et al., 1987).

In the same way that other membranes are subject to fouling, which happens when impurities and pollutants build on the surface of the membrane, these membranes are also susceptible to fouling (Gtari et al., 2023). The reversal of polarity in the ED system serves to alleviate fouling by effectively eliminating foulants from the membrane surface. The EDR technology, initially devised by Ionics in 1974, introduced a modified approach that obviated the necessity of including acid or anti-scalant compounds in ED units (Bayod-Rújula & Martínez-Gracia, 2009; Mir & Bicer, 2021).

6.3.1 ED/RO Hybrid System

Thampy et al. (2011) have developed domestic water desalination and purifying systems based on RO. The recovery of freshwater from these systems typically ranges from 10% to 20%, depending upon the TDS in the water supply (Jarin et al., 2023). The issue of drinking water salinity is escalating due to the loss of groundwater resources. The inadequate recovery of freshwater water over desalination presents a significant issue, particularly in water scarcity (Mir & Bicer, 2021). This publication presents the ED-RO hybrid technology to recover the produced water from BW while maintaining water quality (Hamdan et al., 2021). A high TDS environment (low system resistance leading to high efficacy) was the condition in which the ED unit operated in the hybrid process. The RO system was most effective in low TDS environments to mitigate the membrane's salt burden. Investigators separately tested ED and RO units for households to desalinate water. When you want to desalinate BW with a TDS of 2000–4000 parts per million (ppm), performing the work requires more than just a modest ED unit or RO system (Bian & Jun, 2017). An ED-RO hybrid desalination system was designed to mitigate significant water loss over desalination. The disclosed hybrid technique enables the recovery of more significant quantities of water. The current invention presents an alternate method for achieving over 50% recovery of high-quality product water, even while utilizing BW with a concentration of 4000 ppm as feed. The described hybrid technique is shown to be an optimal home water filtration system for desalination with due diligence (Bian et al., 2019).

Bahrain, which has solar power and 13 hours of sunshine in the summer and 10 hours in the winter, was the site for investigation by HMN AlMadani (2003). Using PV systems, an ED procedure can be carried out to desalinate water efficiently using solar energy. This method is attractive because ED relies on a direct current power supply to drive the removal of salt ions. A commercially available, smaller-scale ED device powered by solar cells was used in this experimental work. It has two electrical phases and four hydraulic phases, with a total of 24 cell pairs.

Two sources supplied water: laboratory-synthesized sodium chloride solutions and moderately salinized groundwater. The product flow rates ranged from 50 to 300 gallons/day, while the temperature ranged from 10°C to 40°C. Augmenting the flow rate led to a diminished product quality, as measured by the percentage of salt removal. Elevating the temperature typically leads to enhanced product quality. PV cells were successfully employed to desalinate brackish fluids, including laboratory-prepared aqueous sodium chloride solutions and naturally occurring subsurface water of moderate salinity. The overall quality of the diluted output was significantly enhanced at low flow rates. Salt reduction reached 95.0% for underground water and 99.0% for sodium chloride solutions at modest product flow rates of 150.0 gallons/day. Elevating the feed water temperature leads to an increased percentage of salt removal at elevated feed flow rates.

Uche et al. (2013) explored the feasibility of a PV-integrated ED unit in Northern Chile due to its high salinity levels and sun radiation, which make it an attractive option for solar-powered energy storage. When coupled with 8 PV panels, the ED approach produced 0.289 $m^3 h^{-1}$ heat in under 0.35 hours. The capacity of an ED plant decreased to 0.22 $m^3 h^{-1}$, and the duration increased to 0.44 h when PV panels were reduced to 4. The plant efficiently produced water with salinity below 1000 ppm and reached a capacity of 1000 $Lday^{-1}$. The largest PV hybrid energy conversion system was installed in Japan in 1990. With a power consumption range of 0.6 to 1 kWh/m^3, the system can produce 200 m^3/day. Wright and Winter (2014) designed a PV-ED system suitable for a community for efficient and economical groundwater desalination. Unlike fossil fuel-powered EDS, a PV-integrated ED system optimized for BW can reduce CO_2 emissions by 0.723 $kgCO_2$-eq. m^{-3}.

An investigation on RO-ED systems for desalinating brackish and extremely salty brine waters was carried out by McGovern et al. (2014). Incorporating RO into the system can reduce the expenditure of ED in desalinating BW. With a recovery rate of 50%, the RO unit processes feed streams with a TDS of 3000 ppm in a single stage of operation. The ED is fed with the RO concentrate stream, which has a TDS of 6000 ppm. They found two advantages to a hybrid system. First, it enables larger current densities and, by extension, higher salt removal rates at increasing salinities, reducing the total membrane area. Second, since the end product is a mixture of ED and RO permeates, the ED permeate can have a greater salinity over the water in the end product, which is an advantage. While the most significant cost reduction is seen when salt removal occurs at high salinity levels, the most significant decrease in cost is seen when both the supply and product salinities are low. If the demand for drinking water is the same as the demand for product water, with a target of 500 ppm TDS, then hybridization would be justified if the relative cost of a single-stage RO system was 65% of that of a standalone ED system (Chekifi et al., 2024). The possible savings from employing a hybrid system are further reduced if the standards for product water are less strict.

Thampy et al. (2011) developed a hybrid process integrating ED and RO to desalinate BW. The approach achieved a high recovery rate, where the permeate from the ED process was used as the feed for the RO process. ED technology is a pretreatment method in high salinity zones, where optimal solution conductivity leads to enhanced efficiency. The RO unit functions within low salinity ecosystems, characterized by low osmotic pressure and, thus, a reduced energy demand for RO processes. Water was generated for every feed TDS range of 2000–4000 ppm with more than 50% recoveries and 50–120 ppm TDS levels. The laboratory-scale experimental ED system was ready for action with an effective membrane area of 200.0 cm^2 and 30 cell pairs. It requires 7.7 kWh/m^3 to produce water with a 10% to 20% recovery rate using RO alone. The hybrid ED-RO system frequently achieved a much greater recovery rate of 50%–60% while consuming 8–10 kWh/m^3. The recovery rate of the hybrid ED-RO system is three to five times higher than that of solo RO, although its energy consumption is about the same or slightly higher than RO alone. Such high recovery rates are impossible with isolated RO at similar SEC. A similar approach was put out by Galama et al., who desalinated seawater using electric discharge (ED) as a pretreatment step.

Zhang et al. (2010) evaluate the efficacy of incorporating ED into RO concentrate inside a modified UF-RO system intended for the treatment of secondary sewage from a wastewater treatment facility. The ED system demonstrated an overall recovery rate of 95%, producing effluent quality similar to the system without RO, albeit with a higher

TOC content. An ozonation method efficiently resolved the issue, which reduced the accumulation of organic compounds. Nonetheless, the current study did not incorporate any data regarding energy consumption.

Turek et al. (2012) compared four different layouts utilized in saltwater desalination facilities. The configurations included a single-stage standalone RO system, a hybrid ED-RO system, an NF-SWRO system, and an NF-SWRO-ED system. A comparison analysis was performed on the SEC and recovery rates of the four systems. With an overall energy input of 6.89 kWh/m^3, the hybrid NF-SWRO-ED system achieved a considerable percentage recovery rate of 69%. The single-stage RO system obtains a recovery rate of about 43%, but the independent RO system has an SEC of 2.75 kWh/m^3, which is significantly lower. In addition, the findings of the research underline that the saturated brine created by the system has the potential to be utilized in the manufacturing of salt as well as in the chlor-alkali production industry.

Ahdab et al. (2020) examine the cost-effectiveness of employing monovalent selective electrodialysis (MSED) to desalinate seawater in greenhouses, particularly in comparison to the commonly utilized RO process. In contrast to RO, which processes detrimental monovalent ions and advantageous divalent ions, membrane sulfate electrolysis exhibits selectivity by exclusively eliminating monovalent ions while preserving the helpful nutrients. A comparative analysis assesses the selectivity, limiting current, and resistance of the Neosepta and Fujifilm MSED membranes. A cost model compares the operational and capital expenses associated with MSED and RO, considering fertilizer and water conservation variables. The results indicate that although MSED is typically more costly than RO, it can be economically viable in specific circumstances, especially when considering reduced electricity expenses and developments in membrane technology. This study additionally investigates the viability of solar-powered desalination as a sustainable option.

A. Galama et al. (2014) examined the feasibility of ED for saltwater desalination and evaluated the energy losses involved in the process. The integration of ED with brackish water reverse osmosis (BWRO) is proposed as a substitute desalination method for SWRO. Experiments have been conducted using a recycled batch ED. As a result, they estimate that the membrane stack primarily accounts for the principal energy loss within the system. Energy losses during water transport are usually rather small. Electroosmosis is directly proportional to the supplied current density, although osmotic water movement is substantial at low supplied current densities, leading to substantial energy loss. Although the percentage energy loss owing to back diffusion became more apparent at a minimum provided current density of 10 A/m^2, it was found to be of little relevance at higher current densities. When the current density is less than 50.0 A/m^2, a hybrid system that combines ED with BW RO does not reduce energy consumption compared to ED alone. ED provides desalination at a lower energy cost and decreased operational expenses than SWRO at a low applied current density of 10.0 A/m^2.

He et al. (2024) created a time-varying EDR method that adeptly utilizes abundant solar power for desalination. This demonstrated that PV-powered desalination technology could adjust pumping and EDR power to align with the supply of intermittent solar energy, hence optimizing the desalination process. Initial results indicate enhanced system performance by directly utilizing 77% of the accessible solar energy—91% greater than traditional systems—and a 92% decrease in battery dependence. A village-scale desalination investigation in India demonstrated that these system enhancements resulted in a 22% decrease in water costs, rendering the technology comparable with existing on-grid, village-scale RO systems primarily fueled by fossil fuels. Future innovations may further decrease expenses, offering an enhanced, sustainable resolution to water scarcity in isolated regions.

The flexible energy demand response system constantly regulated its voltage and flow rate following the varying availability of solar power. In contrast, a standard EDR system maintained a consistent voltage and flow rate (Hussain et al., 2022). The flexible EDR system exhibited a notable enhancement in the utilization of solar energy, effectively harnessing 75% of the accessible solar energy in contrast to the standard system's 40% yield. Therefore, there was a significant reduction of 92% in battery dependence. The flexible EDR batches exhibited an SEC of 3.3 kWh/m^3, almost 40% greater than the SEC of the constant EDR batches, which measured 2.4 kWh/m^3. However, the adaptive system exhibited a higher water production rate relative to the required capital investment in solar power, resulting in a decrease in the LCOW (Elfaqih et al., 2024). The inherent capacity of the flexible EDR system to synchronize its electrical power use with the availability of solar electricity significantly improves its efficiency and cost-effectiveness. The system's ability to minimize reliance on batteries reduces capital and maintenance expenses, rendering it a feasible alternative for remote regions characterized by variable grid electrical availability (Wright & Winter, 2014).

The study by Xu et al. (2020) investigates the effectiveness of a compact solar direct-power ED device utilized for desalination. The study examines the influence of voltage, flow rate, and electrolysis compartment dimensions on the overall efficacy of the desalination process. The experimental setup consists of solar controllers, PV modules, an ED compartment, newly sourced water tanks, valves, untreated water tanks, and flow measurement devices. Through the use of a flow meter and valve, the ED unit can be adjusted to regulate the rate of hydrolysis in both the saltwater tank and the freshwater storage vessel. Both ammeter and voltmeter are used to monitor and evaluate the activities of the PV system, which supplies energy to the ED unit. According to the experimental data, the salinity decreases proportionally at the ED unit's output when the voltage increases. The salinity ranges from 17.03 mol/m^3 at an electrical potential of 30 V to 36.03 mol/m^3 at its highest. With a voltage of 20 V, the salinity drops to 19.13 mol/m^3 at its lowest point and rises to 38.4 mol/m^3 at its peak. With a 10 V voltage, the lowest salinity detected is 22.4 mol/m^3, and the highest is 39.5 mol/m^3. The inquiry also examines the influence of weather conditions on the system's operational effectiveness. Under optimal sunlight conditions, the PV panel produces around 30 V, allowing the salinity at the outlet to decrease below the set threshold (1000 ppm) 50 minutes and 20 minutes sooner than in overcast and cloudy conditions, respectively. During 120 minutes, the total amount of freshwater produced was measured at flow rates of 200.0 mL/h, 500.0 mL/h, 800.0 mL/h, and 1100.0 mL/h. The produced quantities were 0.4, 1, 1.6, and 2.2 L, respectively.

Fernandez-Gonzalez et al. (2015) examine the long-term viability of an ED that utilizes PV solar energy to treat BW. The experimental configuration comprises an ED unit situated in the Canary Islands, designed specifically to address the treatment of BW containing TDS within the range of 2000 to 5000 mg/L. The plant's daily water production is below 1000 m^3, and its objective is to generate freshwater with a TDS concentration of 200 mg/L. The study reveals that using PV power in ED (Environmental Energy) reduces BW's energy usage from 0.49 to 0.91 kWh/m^3 and CO_2 emissions from 0.02 to 0.03 kg/m^3. This significantly reduces the environmental footprint compared to traditional grid-powered ED. The study predicts that the production cost of freshwater using ED-PV will be between 0.15 and 0.4 €/m^3 by 2020.

In their study, Campione et al. (2020) presented a methodology for dynamic simulation and operation of an ED desalination unit that incorporates solar and wind electricity. Flow rates ranging from 230.0 to 920.0 m^3/day and energy supplies ranging from 5 to 45 kW were used to analyze dynamic scenarios. A concentration of 0.24 g/L of NaCl was determined for the outflow. The plant ran most of the year with a combined operational lifetime of 5500 hours, occasionally turning off in winter when irradiation levels were low. According to simulations of their output fluctuations, energy dispatch works well with variable power sources like wind and PV power. This system can handle product flows ranging from tens to thousands of m^3/h, making it suitable for small-scale and industrial local applications. Based on the findings, this polygenerational technique does not require batteries to provide potable water.

Electrically powered water desalination systems were developed by DG Oré et al. (2022) for use at different scales. It is the goal of this study to determine the SEC and the percentage of salinity removal (PSR) under various ED system conditions, as well as to identify the various components of an autonomous photovoltaic system (APVS) that can continuously power the UF-ED device in a laboratory condition. A pair of PV modules and a battery were selected. An output of 150.0 Wp was chosen for every solar module, in addition to a battery capacity of 200.0 Ah and 12.0 V. A charging controller with a power rating of more than 300 Wp, an ideal charge current greater than 18.5 A, and an input voltage greater than 21.6 V was elected. This inverter was selected because its output power is greater than 112.0 W. Multiple operating conditions were used to evaluate the PSR and SEC: 6.0 V at 25.0 mL/min, 12.0 V at 25.0 mL/min, 6.0 V at 50.0 mL/min, and 12.0 V at 50.0 mL/min. Using 12.0V and 25.0 mL/min, the UF-ED system consistently reduced groundwater salinity by 64%.

6.4 CONCLUSION

PV-powered desalination systems offer a sustainable and cost-effective alternative to traditional methods based on fossil fuels. These systems use solar energy to generate direct current electricity, which can be used immediately or stored for future use. Integrating PV with RO and ED technologies enhances energy efficiency and reduces greenhouse gas emissions. These systems are particularly effective for seawater desalination and suitable for BW with lower salinity. Case studies and experimental investigations have shown the technical and economic viability of PV-powered desalination systems in countries like South Africa, India, Greece, and Iran. ACCUs for energy storage ensure continuous operation even during low solar irradiation. Hybrid systems that combine RO and ED technologies optimize desalination processes and reduce energy consumption. This approach offers a sustainable and efficient solution for providing clean drinking water in regions with limited access to freshwater resources.

ABBREVIATION

CEM	Cation-exchange Membranes
ED	Electrodialysis
TDS	Total Dissolved Solids
AEM	Anion Exchange Membranes
PV	Photovoltaic
ED-PV	Electrodialysis-reverse Osmosis
NF-SWRO	Nanofiltration-seawater Reverse Osmosis
SEC	Specific Energy Consumption
NF-SWRO-ED	Nanofiltration-reverse Osmosis-electrodialysis
BWRO	Brackish Water Reverse Osmosis

REFERENCES

Ahdab, Y. D., Schücking, G., Rehman, D., & Lienhard, J. H. (2021). Cost effectiveness of conventionally and solar powered monovalent selective electrodialysis for seawater desalination in greenhouses. *Applied Energy*, *301* (November). https://doi.org/10.1016/j.apenergy.2021.117425

Ahdab, Y. et al. (2020). Brackish water desalination for greenhouses: Improving groundwater quality for irrigation using monovalent selective electrodialysis reversal. *Journal of Membrane Science*. https://www.sciencedirect.com/science/article/pii/S0376738820306517

Al-Amshawee, S., Yunus, M. Y. B. M., Azoddein, A. A. M., Hassell, D. G., Dakhil, I. H., & Hasan, H. A. (2020). Electrodialysis desalination for water and wastewater: A review. *Chemical Engineering Journal*, *380* (March 2019). https://doi.org/10.1016/j.cej.2019.122231

Al-Karaghouli, A., Renne, D., & Kazmerski, L. L. (2010). Technical and economic assessment of photovoltaic-driven desalination systems. *Renewable Energy*, *35*(2), 323–328. https://doi.org/10.1016/j.renene.2009.05.018

AlMadani, H.M.N. (2003). Water desalination by solar powered electrodialysis process. *Renewable Energy*. https://www.sciencedirect.com/science/article/pii/S0960148103000144

Al-Sahali, M., & Ettouney, H. (2007). Developments in thermal desalination processes: Design, energy, and costing aspects. *Desalination*, *214*(1–3), 227–240. https://doi.org/10.1016/j.desal.2006.08.020

Andersson, S. (2009). Characterization of Bacterial Biofilms for Wastewater Treatment. *Technology, 71*. http://kth.diva-portal.org/smash/get/diva2:209486/FULLTEXT01

Association, A. W. W. (1999). Electrodialysis and Electrodialysis Reversal. In *AWWA Manual M38* (Issue C, p. 72).

Bayod-Rújula, A. A., & Martínez-Gracia, A. (2009). Photovoltaic system for brackish water desalination by electrodialysis and electricity generation. *Desalination and Water Treatment, 7*(1–3), 142–151. https://doi.org/10.5004/DWT.2009.717

Belessiotis, V., Kalogirou, S., & Delyannis, E. (2016). Thermal Solar Desalination: Methods and Systems. In *Thermal Solar Desalination: Methods and Systems* (pp. 1–368). https://doi.org/10.1016/C2015-0-05735-5

Bermudez-Contreras, A., Thomson, M., & Infield, D. G. (2008). Renewable energy powered desalination in Baja California Sur, Mexico. *Desalination, 220*(1–3), 431–440. https://doi.org/10.1016/j.desal.2007.01.046

Bian, D. W., & Jun, C. (2017). *Design, development, and field-testing of a cost-optimized village-scale, photovoltaic-powered, electrodialysis reversal water desalination system for rural India*. https://dspace.mit.edu/handle/1721.1/111773

Bian, D. W., Watson, S. M., Wright, N. C., Shah, S. R., Buonassisi, T., Ramanujan, D., Peters, I. M., & Winter, A. G. (2019). Optimization and design of a low-cost, village-scale, photovoltaic-powered, electrodialysis reversal desalination system for rural India. *Desalination, 452* (August 2018), 265–278. https://doi.org/10.1016/j.desal.2018.09.004

Campione, A., Cipollina, A., Calise, F., Tamburini, A., Galluzzo, M., & Micale, G. (2020). Coupling electrodialysis desalination with photovoltaic and wind energy systems for energy storage: Dynamic simulations and control strategy. *Energy Conversion and Management, 216*, 112940. https://doi.org/10.1016/j.enconman.2020.112940

Chekifi, T., Benmoussa, A., & Boukraa, M. (2024). Desalination Powered by Renewables: A Challenge and an AI Opportunity. In *Water Resources Management* (Issue 0123456789). Springer Netherlands. https://doi.org/10.1007/s11269-024-03935-4

Clarke, D., Al-Abdeli, Y. et al. (2013). The effects of including intricacies in the modelling of a small-scale solar-PV reverse osmosis desalination system. *Desalination*. https://www.sciencedirect.com/science/article/pii/S0011916412006017

Darwish, M. A., Abdulrahim, H. K., Hassan, A. S., & Mabrouk, A. A. (2016). PV and CSP solar technologies & desalination: economic analysis. *Desalination and Water Treatment, 57*(36), 16679–16702. https://doi.org/10.1080/19443994.2015.1084533

El Haj Assad, M., Ahmadi, M. H., Sadeghzadeh, M., Yassin, A., & Issakhov, A. (2021). Renewable hybrid energy systems using geothermal energy: hybrid solar thermal–geothermal power plant. *International Journal of Low-Carbon Technologies, 16*(2), 518–530. https://doi.org/10.1093/ijlct/ctaa084

Elfaqih, A. K., Elbaz, A., & Akash, Y. M. (2024). A review of solar photovoltaic-powered water desalination technologies. *Sustainable Water Resources Management, 10*(3), 1–20. https://doi.org/10.1007/s40899-024-01067-6

Ettouney, H., & Wilf, M. (2009). Conventional Thermal Process. *Seawater Desalination: Conventional and Renewable Energy Processes, 306*. https://doi.org/10.1007/978-3-642-01150-4

Fernandez-Gonzalez, C., Dominguez-Ramos, A., Ibañez, R., & Irabien, A. (2015). Sustainability assessment of electrodialysis powered by photovoltaic solar energy for freshwater production. *Renewable and Sustainable Energy Reviews, 47*, 604–615. https://doi.org/10.1016/j.rser.2015.03.018

Freire-Gormaly, M., Bilton, A.M. (2019a). Design of photovoltaic powered reverse osmosis desalination systems considering membrane fouling caused by intermittent operation. *Renewable Energy*. https://www.sciencedirect.com/science/article/pii/S0960148118313843

Freire-Gormaly, M., Bilton, A.M. (2019b). Impact of intermittent operation on reverse osmosis membrane fouling for brackish groundwater desalination systems. *Journal of Membrane Science*. https://www.sciencedirect.com/science/article/pii/S0376738818332447

Galama, A. H., Saakes, M., Bruning, H., Rijnaarts, H. H. M., & Post, J. W. (2014). Seawater predesalination with electrodialysis. *Desalination*, *342*, 61–69. https://doi.org/10.1016/j.desal.2013.07.012

Galama, A., Saakes, M., Bruning, H. et al. (2014). Seawater predesalination with electrodialysis. *Desalination*. https://www.sciencedirect.com/science/article/pii/S0011916413003342

Garg, M. C., & Joshi, H. (2015). A Review on PV-RO Process: Solution to Drinking Water Scarcity due to High Salinity in Non-Electrified Rural Areas. In *Separation Science and Technology (Philadelphia)* (Vol. 50, Issue 8). https://doi.org/10.1080/01496395.2014.951725

Gtari, O., Hamdi, M., & Hazami, M. (2023). Energy behaviour and economic analysis of a photovoltaic-thermal (PV/T) collector coupled with a reverse osmosis (RO) desalination unit in Tunisian climatic conditions: a feasibility study. *Environmental Science and Pollution Research*, *30*(32), 79120–79135. https://doi.org/10.1007/s11356-023-27929-8

Hamdan, H., Saidy, M., Alameddine, I., & Al-Hindi, M. (2021). The feasibility of solar-powered small-scale brackish water desalination units in a coastal aquifer prone to saltwater intrusion: A comparison between electrodialysis reversal and reverse osmosis. *Journal of Environmental Management*, *290*(April 2021), 112604. https://doi.org/10.1016/j.jenvman.2021.112604

He, W., Le Henaff, A.-C., Amrose, S., Buonassisi, T., Peters, I. M., & Winter, A. G. (2024). Flexible batch electrodialysis for low-cost solar-powered brackish water desalination. *Nature Water*, *2*(4), 370–379. https://doi.org/10.1038/s44221-024-00213-w

Herrero-Gonzalez, M., Culcasi, A., Tamburini, A., Ibañez, R., Cipollina, A., & Micale, G. (2024). Techno-economic feasibility of photovoltaic solar electrodialysis with bipolar membranes. *Desalination*, *582*(November 2023). https://doi.org/10.1016/j.desal.2024.117624

Herrero-Gonzalez, M., Diaz-Guridi, P., Dominguez-Ramos, A., Ibañez, R., & Irabien, A. (2018). Photovoltaic solar electrodialysis with bipolar membranes. *Desalination*, *433*, 155–163. https://doi.org/10.1016/j.desal.2018.01.015

Hussain, A., Yan, H., Ul Afsar, N., Jiang, C., Wang, Y., & Xu, T. (2022). Multistage-batch bipolar membrane electrodialysis for base production from high-salinity wastewater. *Frontiers of Chemical Science and Engineering*, *16*(5), 764–773. https://doi.org/10.1007/s11705-021-2114-2

Jacob, J., Pandey, A., Rahim, N. A. et al. (2022). Concentrated Photovoltaic Thermal (CPVT) systems: Recent advancements in clean energy applications, thermal management and storage. *Journal of Energy Storage*. https://www.sciencedirect.com/science/article/pii/S2352152X21010604

Jarin, M., Dou, Z., Gao, H., Chen, Y., & Xie, X. (2023). Salinity exchange between seawater/brackish water and domestic wastewater through electrodialysis for potable water. *Frontiers of Environmental Science and Engineering*, *17*(2), 1–10. https://doi.org/10.1007/s11783-023-1616-1

Jeff, M. (2019). Why Agriculture's Greenhouse Gas Emissions Are Almost Always Underestimated. In *Green Tech*. Green Tech.

Joyce, A., Loureiro, D., Rodrigues, C., Desalination, S. C.-, & 2001, undefined. (2001). Small reverse osmosis units using PV systems for water purification in rural places. *Elsevier*. https://www.sciencedirect.com/science/article/pii/S0011916401002028

Kalidasan, B., Divyabharathi, R., Pandey, A. K., Subramaniyan, C., & Mohankumar, S. (2021). Technological Advancement of Solar Thermal System Desalination Process-A Review. *IOP Conference Series: Materials Science and Engineering, 1059*(1). https://doi.org/10.1088/1757-899X/1059/1/012061

Karavas, C. S., Arvanitis, K. G., & Papadakis, G. (2019). Optimal technical and economic configuration of photovoltaic powered reverse osmosis desalination systems operating in autonomous mode. *Desalination, 466*(December 2018), 97–106. https://doi.org/10.1016/j.desal.2019.05.007

Karimi, L., Abkar, L., Aghajani, M., & Ghassemi, A. (2015). Technical feasibility comparison of off-grid PV-EDR and PV-RO desalination systems via their energy consumption. *Separation and Purification Technology, 151*, 82–94. https://doi.org/10.1016/j.seppur.2015.07.023

Kumar, L., Soomro, J., Khoharo, H., & El Haj Assad, M. (2023). A comprehensive review of solar thermal desalination technologies for freshwater production. *AIMS Energy, 11*(2), 293–318. https://doi.org/10.3934/ENERGY.2023016

Kuroda, O., Takahashi, S., Wakamatsu, K., Itoh, S., Kubota, S., Kikuchi, K., Eguchi, Y., Ikenaga, Y., Sohma, N., & Nishinoiri, K. (1987). An electrodialysis sea water desalination system powered by photovoltaic cells. *Desalination, 65*(C), 161–169. https://doi.org/10.1016/0011-9164(87)90132-9

Lotfy, H. R., Staš, J., & Roubík, H. (2022). Renewable energy powered membrane desalination — review of recent development. *Environmental Science and Pollution Research, 29*(31), 46552–46568. https://doi.org/10.1007/s11356-022-20480-y

Luqman, M., Bicer, Y., & Al-Ansari, T. (2020). Thermodynamic analysis of an oxy-hydrogen combustor supported solar and wind energy-based sustainable polygeneration system for remote locations. *International Journal of Hydrogen Energy, 45*(5), 3470–3483. https://doi.org/10.1016/j.ijhydene.2018.12.191

Mannan, M., & Al-Ghamdi, S. G. (2019). Life-cycle assessment of thermal desalination: Environmental perspective on a vital option for some countries. *World Environmental and Water Resources Congress 2019: Groundwater, Sustainability, Hydro-Climate/Climate Change, and Environmental Engineering - Selected Papers from the World Environmental and Water Resources Congress 2019*, 449–460. https://doi.org/10.1061/9780784482346.046

McGovern, R. et al. (2014). The cost effectiveness of electrodialysis for diverse salinity applications. *Desalination*. https://doi.org/10.1016/j.desal.2014.06.010

Meratizaman, M., & Godarzi, A. A. (2019). Techno-economic feasibility of a solar-powered reverse osmosis desalination system integrated with lithium battery energy storage. *Desalination and Water Treatment, 167*, 57–74. https://doi.org/10.5004/dwt.2019.24550

Mir, N., & Bicer, Y. (2021). Integration of electrodialysis with renewable energy sources for sustainable freshwater production: A review. *Journal of Environmental Management, 289*(April), 112496. https://doi.org/10.1016/j.jenvman.2021.112496

Ndadane, S., Inambao, F. (2022). Consideration of design systems for solar powered reverse osmosis seawater desalination. *Environmental Science, Engineering*. https://www.academia.edu/download/78952162/2_67_1636096317_12IJMPERDDEC202112.pdf

Oré, D. G., Pretell, V. H., & Pilco, A. W. (2022). Design of an autonomous photovoltaic system for desalination of groundwater by electrodialysis. *Journal of Physics: Conference Series, 2180*(1). https://doi.org/10.1088/1742-6596/2180/1/012020

Ortiz, J. M., Expósito, E., Gallud, F., García-García, V., Montiel, V., & Aldaz, A. (2007). Electrodialysis of brackish water powered by photovoltaic energy without batteries: direct connection behaviour. *Desalination, 208*(1–3), 89–100. https://doi.org/10.1016/j.desal.2006.05.026

Ortiz, J. M., Expósito, E., Gallud, F., García-García, V., Montiel, V., & Aldaz, V. A. (2008). Desalination of underground brackish waters using an electrodialysis system powered directly by photovoltaic energy. *Solar Energy Materials and Solar Cells*, *92*(12), 1677–1688. https://doi.org/10.1016/j.solmat.2008.07.020

Panagopoulos, A. (2021). Water-energy nexus: desalination technologies and renewable energy sources. *Environmental Science and Pollution Research*, *28*(17), 21009–21022. https://doi.org/10.1007/s11356-021-13332-8

Petersen, K. L., Frank, H., Paytan, A., & Bar-Zeev, E. (2018). Impacts of Seawater Desalination on Coastal Environments. *Sustainable Desalination Handbook: Plant Selection, Design and Implementation*, 437–463. https://doi.org/10.1016/B978-0-12-809240-8.00011-3

Qasim, M., Badrelzaman, M., Darwish, N. N., Darwish, N. A., & Hilal, N. (2019). Reverse osmosis desalination: A state-of-the-art review. *Desalination*. https://www.sciencedirect.com/science/article/pii/S0011916418325037

Reif, J. H., & Alhalabi, W. (2015). Solar-thermal powered desalination: Its significant challenges and potential. *Renewable and Sustainable Energy Reviews*, *48*, 152–165. https://doi.org/10.1016/j.rser.2015.03.065

Richards, B., & Schäfer, A. (2003). Photovoltaic-powered desalination system for remote Australian communities. *Renewable Energy*. https://www.sciencedirect.com/science/article/pii/S0960148103000818

Riffel, D., & dos Santos Junior, J., & de Moraes Costa, A. (2021). Optimum operational conditions of hybrid photovoltaic-thermal systems. *Energy for Sustainable Development*. https://www.sciencedirect.com/science/article/pii/S0973082620303306

Riffel, D. B., Carvalho, P. C. M., & Riffel, D. B. (2009). Small-scale photovoltaic-powered reverse osmosis plant without batteries: Design and simulation. *Elsevier*, *250*, 42–53. https://doi.org/10.1016/j.desal.2008.07.019

Shatat, M., & Riffat, S. (2014). Water desalination technologies utilizing conventional and renewable energy sources. *International Journal of Low-Carbon Technologies*. https://doi.org/10.1093/ijlct/cts025

Soric, A., Cesaro, R., Perez, P., Guiol, E. et al. (2012). Eausmose project desalination by reverse osmosis and batteryless solar energy: Design for a 1 m3 per day delivery. *Desalination*. https://www.sciencedirect.com/science/article/pii/S0011916412003219

Thampy, S., Desale, G. R., Shahi, V. K., Makwana, B. S., & Ghosh, P. K. (2011). Development of hybrid electrodialysis-reverse osmosis domestic desalination unit for high recovery of product water. *Desalination*, *282*(1418), 104–108. https://doi.org/10.1016/j.desal.2011.08.060

Turek, M., Waś, J., & Mitko, K. (2012). Scaling prediction in electrodialytic desalination. *Desalination and Water Treatment*, *44*(1–3), 255–260. https://doi.org/10.1080/19443994.2012.691747

Uche, J., Círez, F., Bayod, A. et al. (2013). On-grid and off-grid batch-ED (electrodialysis) process: Simulation and experimental tests. *Energy*. https://www.sciencedirect.com/science/article/pii/S0360544213001709

Valero, F., Barceló, A. et al. (2011). Electrodialysis technology-theory and applications. *Desalination*. Retrieved July 18, 2019, from https://www.intechopen.com/download/pdf/13751

Voutetaki, A., Papadopoulos, A., Plakas, K. et al. (2022). Integration of an electrodialysis process for selective nitrate removal with renewable energy sources. *Cetjournal.It*, *94*, 721–726. https://doi.org/10.3303/CET2294120

Wright, N. C., & Winter, A. G. (2014). Justification for community-scale photovoltaic-powered electrodialysis desalination systems for inland rural villages in India. *Desalination*, *352*, 82–91. https://doi.org/10.1016/j.desal.2014.07.035

Xu, H., Ji, X., Wang, L., Huang, J., Han, J., & Wang, Y. (2020). Performance study on a small-scale photovoltaic electrodialysis system for desalination. *Renewable Energy*, *154*, 1008–1013. https://doi.org/10.1016/j.renene.2020.03.066

Zhang, Y., Ghyselbrecht, K., Meesschaert, B., Pinoy, L., & Van Der Bruggen, B. (2010). Electrodialysis on RO concentrate to improve water recovery in wastewater reclamation. *Journal of Membrane Science*, *378*, 101–110. https://doi.org/10.1016/j.memsci.2010.10.036

7 Desalination Batteries and Capacitive Deionization

7.1 INTRODUCTION

The twenty-first century is confronting various significant challenges, among the most crucial of which is the scarcity of safe drinking water (Panagopoulos, 2021). However, conditions that are considered to be underwater-stressed might force around two-thirds of the world's population to relocate by 2025. In 2018, about three billion people were unable to acquire water that was safe to drink. Water technology is fundamental for many industrial processes, including but not limited to mining, crop cultivation, and drinking water supply (Mahato et al., 2022). Several different water purification technologies, including multi-stage flash (MSF), electrodialysis (ED), forward osmosis, and reverse osmosis (RO), have been developed to address this problem (Bhabha Atomic Research Center, 2010). RO is one of the methods utilized extensively for desalinating seawater and brackish water on an industrial scale. It is still an energy-intensive technology that satisfies the massive water demand, even if the cost has decreased significantly from 20 kWh/m^3 to 2-4 kWh/m (Fritzmann et al., 2007).

Blair and Murphy introduced using porous carbon electrodes in electrochemical desalination in 1960 (Suss et al., 2015). This technology is considered to be of the very first generation. Because of the ion electrosorption process, one kind of first-generation cell is called capacitive deionization (CDI) (Z. H. Huang et al., 2017). Each cell has one central channel for the feedwater and two carbon electrodes. According to the concept by Johnson and Newman (1971), ions are drawn out of the water and fixed in the electrical double layer at the fluid-solid interface through electrosorption in the pores of the electrodes when the cell is charged.

DOI: 10.1201/9781003530893-7

CDI is now implemented to remove dissolved, charged components from aqueous solutions (Ahmed & Tewari, 2018). Recently, this technology has been used in the desalination of salty water and seawater, as well as in the remediation of wastewater and water purification (Gao et al., 2019) (Figure 7.1).

Figure 7.1 This schematic diagram illustrates a CDI system featuring a potentiated, CDI cell with an acrylic plate and carbon electrode, peristaltic pump, and conductivity meter linked to a feed water tank. Arrows indicate the flow direction of water.

In recent decades, the area of CDI has experienced significant growth in electrode materials, process analysis, cell designs, and other related aspects (Gao et al., 2019). However, the actual utility of CDI and its continued improvement are still limited by parasitic reactions and a very low capacity for desalination (Porada et al., 2012). The primary side reactions occurring during the charging process in CDI include cathode reduction processes, such as oxygen reduction and anode oxidation reactions; carbon electrode reduction, encompassing carbon electrode oxidation and chloride oxidation; and co-ion expulsion, which entails the removal of ions possessing the same charge as the electrodes (M. A. Ahmed & Tewari, 2018). The occurrence of these reactions regularly during the operation process leads to an increase in the amount of energy lost, which in turn causes fluctuations in the water quality. The modest desalination capacity mainly arises from carbon materials' comparatively low physical charge storage capacitance, which is constrained by their porous structure, electric conductivity, and wettability (Xing et al., 2020).

Additionally, the negative consequences of side reactions further contribute to this limitation. Several approaches have been devised to address these difficulties, such as optimizing the configuration and operation style of the CDI (Tang et al., 2019). One of

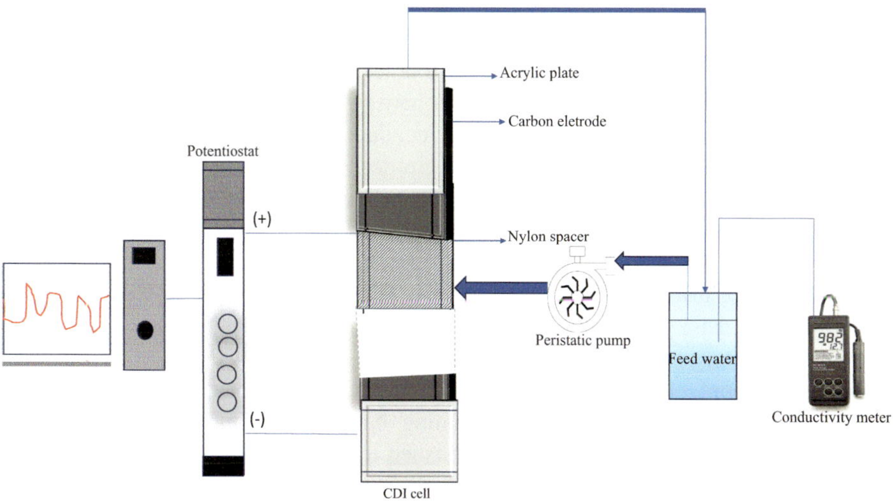

FIGURE 7.1 Schematic diagram of CDI system.

the most effective methods for this is to use faradaic electrode materials for the source electrodes. Faradic electrodes store ions through intercalation or conversion events, as opposed to porous carbon electrodes that rely on capacitance in an electric double layer on surfaces or nanopores (Yu et al., 2019).

In electrochemistry, CDI refers to the storage of energy in the electrodes, followed by the release of the energy by electrode regeneration (Liang et al., 2019). The CDI process is intricately connected to the characteristic behavior of batteries and supercapacitors (Wang, Zhang, Moh, & Presser, 2021). Battery capacities often surpass those of supercapacitors due to the redox processes in the battery electrode materials, facilitating a greater magnitude of charge transfer. The absence of boundaries between battery materials and pseudocapacitive elements can be attributed to their shared foundation in redox processes (Zhang et al., 2018). The development of rechargeable battery materials has shown rapid progress recently, resulting in several types with unique electrochemical characteristics in the literature. Some materials are currently used in manufacturing, suggesting that combining CDI technology and battery materials could be useful (Zhang et al., 2018).

The potential electrochemical capacity of battery materials often exhibits a significant value, potentially enhancing the CDI system's desalination capacity. There are still certain factors that must be considered when choosing electrode materials for battery CDI (Zhang et al., 2018). During desalination, the electrode is submerged in water, so the material needs to be hydrophilic and solid so that the ions can make full contact with the material. To keep the water from splitting, the redox processes must also happen within the voltage range of water. In addition, the ions in the water may hydrate, which makes the measurements greater (Suss et al., 2015). This underscores the necessity for more rigorous structural specifications in battery materials to avert the disintegration of their conformation over both charging and discharging cycles.

Over the past few years, desalination research has primarily focused on incorporating battery materials into the CDI operation (Porada et al., 2013). This chapter includes a comprehensive analysis of the study about battery materials employed as CDI electrodes and their performance metrics. This is conducted to improve our comprehension of the functions executed by battery CDI electrodes. The properties of the starting batteries and the dynamics of CDI operations are considered while categorizing the battery materials employed in the CDI. The materials are subsequently categorized into numerous classes (Yuan et al., 2012).

This chapter investigates the potential of CDI and desalination batteries as innovative ways to increase global water scarcity. The work outlines the benefits of CDI, such as low energy consumption, straightforward operation, and limited secondary contamination, and it emphasizes the role of electrode materials in improving desalination efficiency. The chapter also explores combining battery materials with CDI system, presenting different battery materials established to enhance deionization capabilities. By analyzing the obstacles and developments in CDI and desalination batteries, the chapter aims to provide complete knowledge of these technologies and their ability to contribute to environmentally friendly water purification.

7.2 CAPACITIVE DEIONIZATION

Through the development of desalination technology, the vast quantities of salt water on Earth effectively address the lack of freshwater (Ayaz et al., 2022). Water desalination is accomplished through RO and thermal distillation, the most used commercial procedures. These technologies aren't great for treating low salinity streams like brackish water, with a total dissolved solids (TDS) of about 0.6–3 gL^{-1}, because they use a lot of energy and humans need to remineralize it more (B. Kim et al., 2020). Using CDI technology for brackish water desalination has several advantages, including energy-efficient potential, removing salt ions from the source water at room temperature, low pressure, and voltage, and the potential for energy recovery. This contrasts with traditional desalination methods, which focus on extracting water molecules from salt solutions. Although CDI is currently effective for desalinating brackish water, which can be generated in the laboratory using deionized water (HHitech, China), there is significant potential for enhancement. Many studies have been done to improve the desalination efficiency of brackish water using CDI and to make sure it stays stable since CDI became popular again in the 1990s. Improved electrode material synthesis, new operating mode development, and novel cell design were the three main foci of the research (Martinez et al., 2022).

Soffer and Folman (1972) were the ones who initially presented the concept of monitoring the ability of a cell's charge-discharge cycle to adsorb salt. This method is becoming increasingly used in CDI. The duration of the charge-discharge cycle can vary, ranging from times that are pretty short and involve only a small amount of adsorption to extremely lengthy and involve achieving equilibrium. In order to find the latter, it is necessary to first measure the equilibrium salt adsorption ability, which is also known as the maximal salt adsorption capacity. In order to achieve equilibrium in a CDI salt adsorption experiment, it is imperative to maintain a constant cell voltage until the cell charging process is complete, ensuring that there are no charge gradients remaining in the electrode. This will ensure that the salt concentration is equivalent throughout the cell. Only then can equilibrium be achieved (Suss et al., 2015). In an equilibrium state, the cell effluent's measured conductivity remains constant over time. The amount of salt that is removed from the supply of water can be determined in two ways: either by multiplying the total volume of the solution utilized by the process by the reduction in salt concentration within the framework or by computing the time integral of the differences between the concentrations of the cell intake and effluent, multiplied by the flow rate through the cell from the initiation of the charge until equilibrium is attained.

The advancement of novel electrode materials and their chemical composition modification to improve adsorption capacity and longevity has been a significant focus of study in CDI (Jiang et al., 2020). Approximately 60% of the resources allocated to the domain of CDI have concentrated on enhancing superior electrode materials (Ahmed, Khalil, & Hilal, 2021). The salt adsorption capacity of CDI electrodes composed of diverse single-carbon capacitive materials, including carbon aerogels, activated carbon, carbide-derived carbons, mesoporous carbon, carbon nanofibers and nanotubes, activated carbon cloth, graphenes, and metal-organic framework-derived carbons, was determined to be relatively low. The maximal salt adsorption capacity (SAC) was recorded at under 15 mg g^{-1} (M. A. Ahmed & Tewari, 2018).

The next generation of electrodes will be made out of carbon-based composites, which comprise carbon-metal oxide composites, carbon composites, carbon-polymer-metal oxide, and carbon-polymer composites. Through the utilization of a carbon-coated battery component as a composite electrode, it is feasible to overcome the constraints that are associated with pure battery materials (Jiang et al., 2020). These drawbacks include low electroconductivity and slow ion diffusion rates. This will allow for an even larger salt adsorption capacity. Rather than focusing on increasing salt adsorption capacity, several groups have been concentrating on enhancing electrode materials' conductivity over extended periods. An alternative approach involves applying a chemical treatment to the electrodes, resulting in the anode possessing a net -ve surface charge and the cathode exhibiting a net +ve surface charge (M. A. Ahmed & Tewari, 2018).

The composite's additional components may improve pure carbon desalination by changing the intrinsic pore architecture, mitigating agglomeration, optimizing surface chemistry, facilitating water adhesion, introducing pseudo-capacitance, or modifying the zeta potential. Common metal oxides utilized in carbon compounds include ZrO_2, TiO_2, ZnO, MnO_2, $Fe2O_3$, and Fe_3O_4, CeO_2; commonly utilized polymers in carbon materials include ion-exchange resin, polyaniline, polypyrrole, and chitosan (Jiang et al., 2020).

CDI has received much attention as an alternative to standard RO desalination methods over the past ten years. This is primarily due to its moderate operating conditions, mobility, and potential for minimal energy use. The limited salt-removal ability of capacitive electrodes, averaging 10.0 to 15.0 mg/g, significantly restricts the practical implementation of CDI due to its low desalination efficiency. An operating voltage exceeding 1.2 V is frequently utilized to improve the salt-removal efficiency. This approach enhances side reactions, such as irreversible oxidation of carbon electrodes, the evolution of H_2/O_2 gases, and increased brine concentration. Desalination methods with a high capacity for salt removal are in great demand to address these issues (Jiang et al., 2020).

7.3 DESALINATION BATTERY

An electrolyte made of salt water, an electrode that stores sodium, and an electrode that stores chloride are the main parts of a desalination battery. The desalination battery can remove salt when charging, renew the electrode, and recover energy during discharge when the dielectric potential of the sodium-storage electrode is less compared to the chloride-storage electrode and reversibly (D. Xu et al., 2020).

In 2009, Brogioli introduced a novel approach to harvest and capture the free energy generated by combining two streams featuring varying salt concentrations, such as seawater and river water. The basis of his innovative technology is removing entropic energy and its subsequent storage in an electrochemical double layer (EDL) situated on the electrode's surface. The entropy change that occurs when two water streams with differential salinity are mixed forms the basis for the capacitive mixing (CAPMIX) techniques (Brogioli, 2009). La Mantia et al. (2011) presented a method to address issues related to energy harvesting at the electrode-electrolyte interface by harnessing energy from salinity gradients and storing it as energy of an electrochemical nature contained within the bulk crystal structure of the material that was used for the electrode. This method relates explicitly to battery or faradaic electrodes, wherein energy is stored within chemical bonds. This electrical energy production method is called Mixing Entropy Battery or Battery Mixing (BattMix). Batteries designed to capture energy from water via salinity variations have laid the foundation for an innovative water desalination method. The initial design of a desalination battery (DBs) for seawater desalination was presented and proven in 2012. Unlike CAPMIX or CDI technologies, which accumulate salt at the electrode interface, DBs keep charged ions within the bulk phase. This facilitates enhanced particular desalination rates, reduced energy use, and a more dependable desalination process (Santos & La Mantia, 2023).

The initial database concept comprises a closed cycle established on four sequential stages. The electrodes are initially fully charged, indicating the absence of Cl/Na+ in their bulk crystal structure. They are subsequently immersed in seawater and expelled by applying a continuous stream. This phenomenon facilitates the extraction of ions from the bulk solution and liberates energy (Lu et al., 2024). The deionization/discharge process ends with removing and replacing the freshwater solution with more saltwater. Increasing the battery cell's voltage is the goal of this water exchange procedure. Therefore, the electrodes are charged at this stage, and ions are liberated. Hence, regenerating electrodes/charging is directly linked to producing a concentrated solution by brine. Ultimately, the brine is replaced with saltwater, thus priming the framework for the ensuing deionization cycle. The first setup of this desalination unit featured a cationic sodium insertion electrode ($Na_2xMn_5O_{10}$) and an anionic electrode (Ag/AgCl) for chloride collection (Nurohmah et al., 2022).

7.4 DEFINITION AND CLASSIFICATION OF DESALINATION BATTERIES

7.4.1 Rocking Chair Desalination Battery

In this type of battery, an ion-exchange membrane divides the battery's two parts, including sodium or chloride-storage electrodes arranged bilaterally. Fresh and concentrated water are produced concurrently by capturing ions through the ion-exchange membrane and utilizing ion diffusion during the charging and discharge processes (C. Xu et al., 2021a).

The materials that are used as electrodes for the deinsertion of sodium ions in aqueous electrolytes can be divided into three primary categories such as Prussian Blue Analogues (PBAs), which include nickel hexacyanoferrates and copper; metal oxides, which include V_2O_5, MnO_2, and $Na_{0.44}MnO_2$; and transition/alkali metal phosphates, which include $NaTi_2(PO4)_3$, $NaFePO_4$, $Na_2FeP_2O_7$, or the NASICON-type (T. Kim et al., 2017). Between 2014 and 2016, research was conducted on metal-free hybrid seawater fuel cells and Na and air seawater batteries. To advance the development of rechargeable aqueous sodium batteries, some research was conducted on using sodium chloride as a catholyte. In order to charge the battery, sodium ions obtained from seawater were directed to the (-ve) electrode. This was the operational mode that was dependent on the technique (T. Kim et al., 2017).

The rocking chair DB consists of two chambers filled with NaFeHCF and NaNiHCF, electrically isolated by an Anion Exchange Membrane (AEM). The Na+ ions are extracted from the NaNiHCF cathode and transported to the seawater electrolyte that is contained within the cathode chamber while the battery is being charged. The anion-exchange membrane prevents the entry of cations into the anode chamber (M. A. Ahmed & Tewari, 2018; T. Kim et al., 2017). Because of this, anions must diffuse through the anion-exchange membrane to maintain the charge balance in the solutions. Within the DB, ion transport is responsible for the concentration of solution within the cathode chamber and the dilution of solution within the anode chamber throughout both stages of the charging process. When the discharge operation is being carried out, the reverse procedure takes place. An advantageous feature of this rocking chair DB is its lack of chlorine storage material, which significantly restricts the widespread use of conventional desalination batteries (C. Xu et al., 2021b).

The increasing demand for sodium batteries as a viable substitute for Lithium-based batteries, motivated by their electrochemical similarities (including analogous intercalation chemistry to Lithium) and the existing constraints related to Lithium ions (such as scarcity, limited availability, and consequent high costs of Lithium resources), has led to significant advancements in active materials that can electrochemically store Na+ (Pezeshki & Kettab, 2021).

The oxide has certain disadvantages, one being its relatively low conductivity, which may pose issues in specific designs. The oxide's conductivity is comparatively low, potentially leading to a decrease in the material's durability. This is a disadvantage of the oxide. Integrating carbon into the material is one strategy that may be employed to tackle the issue (Pezeshki & Kettab, 2021). In the framework of the CDI process, Chen et al. (2017) developed the MnO_2/MNCNT composite. The incorporation of carbon nanotubes markedly enhanced the Sodium Rocking-Chair (SRC) by a factor of three, resulting in a 50% augmentation of the electrochemical capacitance. Moreover, the material exhibited an extraordinary recycling ability due to the enhancement of the conductivity of the carbon nanotubes and their specific surface area.

Na-ion desalination (NID) devices provide a foundation for investigating the chloride-ion rocking chair desalination process. In a cation-exchange membrane (CEM), sodium ions migrate across the water to create a charged equilibrium, while chloride ions move between the electrodes to accomplish complete desalination. Like NID cells, the Capacitive Ion Deionization (CID) device has two parts holding the concentrated and diluted streams. A limited number of electroactive chemicals can selectively and reversibly remove Cl ions from water, but NID is one of them. Thus far, most conversion reactions have been reported using Ag/AgCl or Bi-based systems (Yu et al., 2019).

The CDI system designed by Chen et al. (2017) utilized a $Na_{0.44}MnO_2$ electrode for intercalating and deintercalating sodium and an AgCl electrode to maintain the same process for chloride. $Na_{0.44}MnO_2$ significantly reduced the polarization of the electrode since it worked as both a sodium donor and a cathode. On the other hand, AgCl was used as both a chloride donor and an anode material. One of the products of the solid-state reaction was $Na_{0.44}MnO_2$. After charging the system, chloride and sodium were released from the electrodes into the solution. The system discharged chloride and sodium, demonstrating typical battery material behavior. After 100 cycles, it showed exceptional cycling capacity and a higher SRC of 54.3 mg g^{-1}, estimated based on electrode total mass.

Only a few materials that are acceptable for use in aqueous sodium-ion batteries have been investigated. One of these materials is the anode material $NaTi_2(PO4)_3$. $NaTi_2(PO4)_3$ (NTP) is a battery material that is favorable to the environment and has a significant capacity of 133 mAh/g. It also has a voltage plateau of around 0.6 V in comparison to the standard hydrogen electrode (SHE), which is located within the voltage range of water (Pezeshki & Kettab, 2021). Guo et al. (2018) employed a solvothermal method to produce the $NaTi_2(PO_4)_3$/rGO nanocomposite. This nanocomposite exhibited enhanced desalination capabilities and cycling performance relative to pure NTP. The deionization capacities were assessed in a high-capacity deionization system, whereby the nanocomposite functioned as the anode and was enveloped by a CEM. The cathode, including activated carbon, was covered by an AEM. The system functioned in a constant current state. Regarding salt adsorption, the composite material exhibited a capacity of 140.0 mg g^{-1} and maintained a capacity of 120.0 mg g^{-1} after undergoing 100.0 cycles. On the other hand, the pure NTP material had a capacity of 112.0 mg g^{-1} and maintained 71.0 mg g^{-1} throughout approximately 100 cycles. Increasing the material's ionic and electrical conductivity was accomplished by using reduced graphene oxide (Figure 7.2).

The high salt-removal capacity of the NTP material makes it a promising option for forming an energy-efficient CDI system when combined with a high-capacity Ag cathode for chloride ion intercalation. An anode composed of $NaTi_2(PO4)_3$/rGO and a cathode composed of Ag/rGO were utilized to develop a CDI system by Huang and Tang (2019). They achieved a salt-removal capability of 105.0 mg g^{-1} throughout 50 cycles. The total amount of energy that is spent by the system to reduce the salinity of the water from 2500.0 parts per million to 500.0 parts per million is just 0.25 Whg^{-1}, which is equivalent to the amount of energy that is required by the RO approach.

Wang et al. (2018) integrated the desalination procedure with energy storage by creating an aqueous battery that employs a silver cathode, a sodium titanate phosphate anode, and a sodium chloride solution as the electrolyte. The solar cell powered the system, exhibiting a desalination capacity of roughly 151.5 $mg{\cdot}g^{-1}$, with no notable capacity deterioration seen after 50.0 cycles. The battery maintained a 49 $mAh{\cdot}g^{-1}$ capacity after 1500.0 cycles at a current density of 10.0 C, which is 70% of its original capacity. The battery operated at an output voltage of 0.83 V, indicating a positive potential of the device.

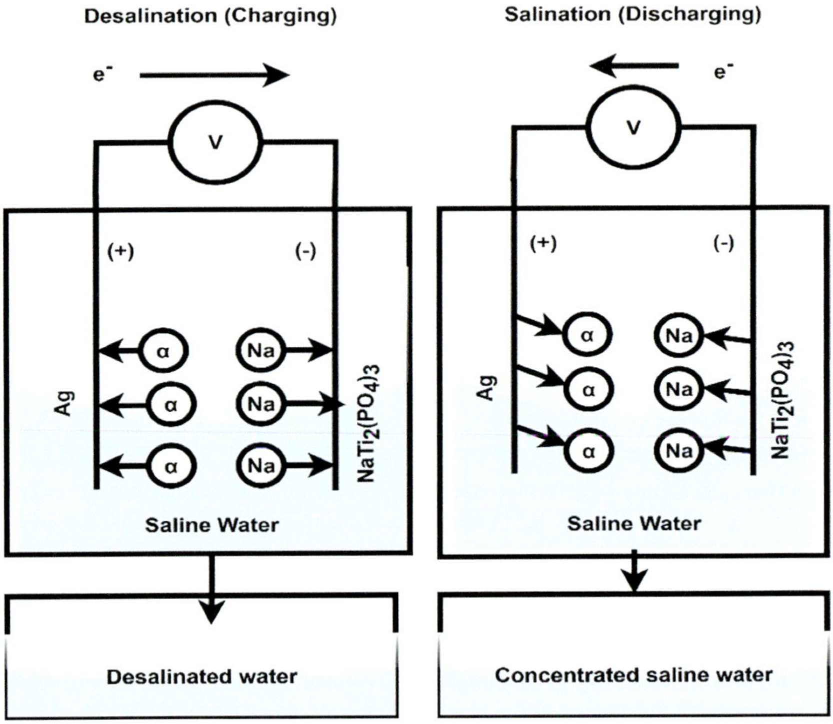

FIGURE 7.2 $NaTi_2(PO_4)_3$||Ag desalination battery.

Kim et al. (2016) presented a CDI system employing $Na_2FeP_2O_7$/C, synthesized through a solid-state method, as the sodium intercalation electrode alongside an activated carbon electrode encased by an anion-exchange membrane. Before the CDI research, the $Na_2FeP_2O_7$ electrode was charged to release sodium ions, generating sodium intercalation sites. In the CDI test, sodium ions were captured by the $Na_2FeP_2O_7$ electrode at a negative voltage, while the activated carbon electrode captured chloride ions. The material exhibited a salt-removal capacity of 32 $mg{\cdot}g^{-1}$ and a specific capacity of 57 $mAh{\cdot}h^{-1}$ at a current density of 48 $mA{\cdot}g^{-1}$ utilizing the galvanostatic method, much exceeding that of traditional carbon electrodes.

7.4.2 Dual-ion Desalination Batteries

A comparison may be drawn between the operational concept of a device that combines the Cl conversion electrodes and the Na+ selective electrodes of a dual-ion battery (Porada et al., 2012). As a result, the concept of capturing and releasing sodium ions and chloride ions in electrodes capable of simultaneously desalinating the feed is conceived, much like a battery. In the technical literature, dual-ion desalination batteries are also referred to as faradaic deionization (FDI) and dual-ion deionization electrochemical deionization (DEDI) (Tang et al., 2019). Dual-ion desalination batteries (DIDBs) are developed to solve CDI technology's limitations linked with its inherent electrosorption mechanism, which uses electrical double layer capacitance (EDLCs). These limitations include that high-concentration solution treatment requires a lot of energy and that ion removal is limited (Guo et al., 2018).

Most designs of DIDB cells incorporate ion-exchange membranes to enhance ion selectivity. Another crucial approach to enhance the operating efficiency of the cell is to ensure a suitable material balance in these devices. This research considers the many theoretical capacities of the electrodes in an aqueous electrolyte. For example, $Na_2Mn_5O_{10}$ (NMO) has a capacity of 40–45 $mAhg^{-1}$, while BiOCl has a capacity of 92 $mAhg^{-1}$ (M. A. Ahmed & Tewari, 2018). The limited sub-surface conversion reaction observed in chloride uptake electrodes is also noted. Approximately 50% of the Bi exhibits electrochemical activity when placed in a 0.6 M sodium chloride solution. Furthermore, it is essential to investigate the expansion and contraction of the conversion electrodes resulting from the phase shift to attain an optimal cell design (Santos & La Mantia, 2023).

7.4.3 Redox-based Desalination Batteries (ReDB)

The capacity to store significant amounts of electrical energy through electrolyte redox connections has increased interest in redox flow batteries (RFBs). Consequently, the RFB cell requires the utilization of two electrolytes, specifically a catholyte and an anolyte, in which the active materials for the cathode and anode are dissolved, accompanied by a separator. These electrolytes facilitate the transportation and storage of energy in distinct reservoirs (Suss et al., 2015). During the charging or discharging process of the battery, the electrolytes are circulated into a stack, enabling the independent scaling of the battery's power and capacity. Redox-active compounds may serve as a DB method for decoupling desalination efficacy from the electrode's capacity. Several research studies have shown that electrochemically driven desalination utilizes redox mediators. However, applying these technologies for desalination necessitates carefully examining various factors, including redefining electrode concepts and alternative design concepts for cells (Lotfy et al., 2022).

7.4.4 Metal-air Desalination (MeDB)

The anode and cathode chambers of a metal-air desalination battery are separate and delineated by anodes and cathodes. The battery uses a metallic anode and an air electrode as cathodic electrodes. It has been demonstrated beyond a reasonable doubt that zinc (Zn) and aluminum (Al) are both viable anodes for saltwater desalination. A zinc-air DB shows the working principle of a metal-air DB to illustrate the process. An air cathode with a constant supply of sodium chloride as the catholyte, a zinc anode with a constant flow of zinc chloride as the anolyte, and a central sodium chloride feed stream comprise the zinc-air DB. During a discharge, the zinc anode goes through the process of oxidation, which removes chloride ions from the salt stream in the middle flow and their subsequent transfer to the anolyte solution (Kim et al., 2017).

The cathodic electrochemical membrane (CM) reduces O_2 in the air to OH– at the cathode. Meanwhile, Na+ ions are absorbed from the intermediate feed stream and transferred to the catholyte stream; this process occurs simultaneously. The charging process releases sodium and chloride ions into the intermediate salt stream. This study demonstrated that it is possible to recover energy from a metal-air battery. A zinc-air DB can illuminate 60 LEDs without needing an external power source (Liang et al., 2019). It is the dissolved oxygen level in the cathode that determines the limitations that are placed on the desalination rate as well as the amount of energy that the battery generates. Additionally, to facilitate the redox process that is carried out at the cathode, a catalyst must be incorporated into the structure of the catholyte (Ahmed & Tewari, 2018).

Figure 7.2 Diagram of a desalination and salination process using a $NaTi_2(PO_4)_3$||Ag desalination battery. Charging (left) shows sodium ions entering $NaTi_2(PO_4)_3$. Discharging (right) shows sodium ions leaving $NaTi_2(PO_4)_3$. Both chambers are connected to the saline water source.

7.5 PROPERTIES OF DESALINATION BATTERIES AND FUTURE APPROACHES

The DB has garnered significant interest as a potential desalination method due to its low energy consumption and excellent salt-removal capacity (Wang, Zhang, Moh, et al., 2021). The DB, however, is a relatively new invention. However, none of the material options that have been reported meet all three requirements: they must be stable over time, work with a range of voltages that doesn't electrolyze water, and be able to remove more than 100 mg of salt per gram of active material (Jiang et al., 2020). As a result, much of the research needs to go into creating and researching Na-storage materials that are easier on the environment and don't cost much capital, like Fe- and Mn-based materials. Future studies should look into this possibility because the transition metals that leak from the Na-storage materials could contaminate the desalinated water in a second way (Cao et al., 2023). Electrode materials coated with carbon and conductive polymers may improve their desalination performance by increasing their conductivity and decreasing the possibility of leaching or dissolving (X. Zhang et al., 2021).

Palladium (PPy) and silver (Ag) are relatively expensive materials. While desalination processes are being carried out, Bi is a hazardous material that causes high-voltage hysteresis. Its low operating potential, which may lead to the release of hydrogen during the desalination process, is another factor that adds to the excessive amount of energy consumed. A novel DB that does not use an electrode for Cl-storage or a material with Cl-storage capabilities is urgently required (Arnold et al., 2022).

When using a DB, it is common practice to place an ion-exchange membrane in front of the electrodes to block the entry of co-ions into the electrodes. This keeps the electrodes safe from salt and water ions (Xu et al., 2020). This is done to improve the desalination system's charging effectiveness and the electrodes' cycle stability. Desalination will become much more expensive due to the costly ion-exchange membrane's limited lifespan and frequent replacement requirements. Therefore, there is an immediate need to develop a DB that does not require membranes (Ahmed, Khalil, et al., 2021).

Also shown is the use of PBAs in water desalination batteries. Because of their easy synthesis and 3D rigid architecture, PBAs have garnered much interest as electrode materials in aqueous batteries. When putting host ions into the PBAs lattice, Na+ is the most suited due to its small hydrated ionic radius and significant energy density among the many options (Anis et al., 2019; Xu et al., 2021a; Xu et al., 2020).

7.6 CONCLUSION

Desalination batteries and CDI is an innovative approach to purifying water in response to the growing water scarcity. The CDI technique is a promising technology with benefits such as easy operation, low energy consumption, and minimum secondary contamination. To improve desalination performance, this chapter gives the significant role that electrode materials offer. The desalination capability of carbon-based materials is rather low, although they are the most frequent. Deionization capabilities have been significantly enhanced by integrating chloride-ion battery materials, including sodium-ion and lithium-ion batteries, into CDI technologies. The issues associated with CDI, including parasitic reactions and limited desalination capacity, are also analyzed. The study examines cell designs and the optimization of electrode materials as potential answers to these difficulties. Creating new types of desalination batteries, including rocking chair and DIDBs, signifies a substantial advancement in the industry. These batteries employ the principles of electrochemical energy storage to accomplish effective desalination. They have the potential to be utilized in the treatment of brown water as well as seawater.

REFERENCES

Ahmed, F. E., Khalil, A., & Hilal, N. (2021). Emerging desalination technologies: Current status, challenges and future trends. *Desalination*, *517*(May), 115183. https://doi.org/10.1016/j.desal.2021.115183

Ahmed, F., Khalil, A. et al. (2021). Emerging desalination technologies: Current status, challenges and future trends. *Desalination*. https://www.sciencedirect.com/science/article/pii/S001191642100254X

Ahmed, M. A., & Tewari, S. (2018). Capacitive deionization: Processes, materials and state of the technology. *Journal of Electroanalytical Chemistry, 813*, 178–192. https://doi.org/10.1016/j.jelechem.2018.02.024

Anis, S., Hashaikeh, R. et al. (2019). Functional materials in desalination: A review. *Desalination*. https://www.sciencedirect.com/science/article/pii/S0011916419311385

Arnold, S., Wang, L., Presser, V., Arnold, S., Wang, L., & Presser, V. (2022). Dual-use of seawater batteries for energy storage and water desalination. *Wiley Online Library, 18*(43). https://doi.org/10.1002/smll.202107913

Ayaz, M., Namazi, M., Din, M. et al. (2022). Sustainable seawater desalination: Current status, environmental implications and future expectations. *Desalination*. https://www.sciencedirect.com/science/article/pii/S0011916422004775

Bhabha Atomic Research Center. (2010). *Desalination & Water Purification Technologies: Technical Information Document*. 40.

Cao, G., Alam, M., Juthi, A., Zhang, Z., et al. (2023). Electro-desalination: State-of-the-art and prospective. *Advanced Membranes*. https://www.sciencedirect.com/science/article/pii/S2772823422000343

Chen, H., Huang, Z., Huang, Y., Zhang, Y., Ge, Z., Qin, B., Liu, Z., Shi, Q., Xiao, P., Yang, Y., Zhang, T., & Chen, Y. (2017). Synergistically assembled MWCNT/graphene foam with highly efficient microwave absorption in both C and X bands. *Carbon*. https://doi.org/10.1016/j.carbon.2017.09.007

Brogioli, D. (2009). Extracting renewable energy from a salinity difference using a capacitor. *APSD Brogioli Physical Review Letters, 2009•APS, 103*(5). https://doi.org/10.1103/PhysRevLett.103.058501

Fritzmann, C., Löwenberg, J., Wintgens, T., & Melin, T. (2007). State-of-the-art of reverse osmosis desalination. *Desalination, 216*(1–3), 1–76. https://doi.org/10.1016/j.desal.2006.12.009

Gao, X., Omosebi, A., Ma, Z., Zhu, F., Landon, J., Ghorbanian, M., Kern, N., & Liu, K. (2019). Capacitive deionization using symmetric carbon electrode pairs. *Environmental Science: Water Research and Technology, 5*(4), 660–671. https://doi.org/10.1039/c8ew00957k

Guo, L., Huang, Y., Ding, M. et al. (2018). A high performance electrochemical deionization method to desalinate brackish water with an FePO 4/RGO nanocomposite. *Journal of Materials Chemistry A*. https://pubs.rsc.org/en/content/articlehtml/2018/ta/c8ta01361f

Huang, K., & Tang, H. (2019). Temperature and desorption mode matter in capacitive deionization process for water desalination. *Environmental Technology, 41*(26), 3456–3463. https://doi.org/10.1080/09593330.2019.1611941

Huang, Z. H., Yang, Z., Kang, F., & Inagaki, M. (2017). Carbon electrodes for capacitive deionization. *Journal of Materials Chemistry A, 5*(2), 470–496. https://doi.org/10.1039/c6ta06733f

Jiang, Y., Alhassan, S. I., Wei, D., & Wang, H. (2020). A review of battery materials as cdi electrodes for desalination. *Water (Switzerland), 12*(11), 1–32. https://doi.org/10.3390/w12113030

Johnson, A. M., & Newman, J. (1971). Desalting by means of porous carbon electrodes. *Journal of The Electrochemical Society, 118*(3), 510–517. https://doi.org/10.1149/1.2408094/META

Kim, B., Seo, J. Y., & Chung, C. H. (2020). A hybrid system of capacitive deionization and redox flow battery for continuous desalination and energy storage. *Journal of Power Sources, 448* (June), 227384. https://doi.org/10.1016/j.jpowsour.2019.227384

Kim, S., Lee, J., Kim, C., et al. (2016). Na2FeP2O7 as a novel material for hybrid capacitive deionization. *Electrochimica Acta*. https://www.sciencedirect.com/science/article/pii/S0013468616308556

Kim, T., Gorski, C. A., & Logan, B. E. (2017). Low Energy Desalination Using Battery Electrode Deionization. *Environmental Science and Technology Letters, 4*(10), 444–449. https://doi.org/10.1021/acs.estlett.7b00392

Liang, Q., Chen, F., Wang, S., Ru, Q., He, Q., Hou, X., Su, C. Y., & Shi, Y. (2019). An organic flow desalination battery. *Energy Storage Materials*, *20*, 203–207. https://doi.org/10.1016/J.ENSM.2018.11.006

Lotfy, H. R., Staš, J., & Roubík, H. (2022). Renewable energy powered membrane desalination — review of recent development. *Environmental Science and Pollution Research*, *29*(31), 46552–46568. https://doi.org/10.1007/s11356-022-20480-y

Lu, Y., Yuan, Y., Liang, T., Jia, H., Meng, L., Zhang, X., Zhang, S., Wen, B., Feng, Z., Yin, T., Guan, P., Zhou, L., Zhou, Y., & Chu, D. (2024). Mechanisms elucidation of secondary seawater batteries: From ion migration to conversion for sustainable energy storage. *Chemical Engineering Journal*, *498*(August), 155307. https://doi.org/10.1016/j.cej.2024.155307

Mahato, A., Upadhyay, S., & Sharma, D. (2022). Global water scarcity due to climate change and its conservation strategies with special reference to India: A review. *Plant Archives*, 64–69. https://doi.org/10.51470/PLANTARCHIVES.2022.V22.NO1.009

Mantia, F. La, Pasta, M., Deshazer, H. D., Logan, B. E., & Cui, Y. (2011). Batteries for efficient energy extraction from a water salinity difference. *ACS Publications*, *11*(4), 1810–1813. https://doi.org/10.1021/nl200500s

Martinez, J., Colán, M., Catillón, R., Huamán, J., Paria, R., Sánchez, L., & Rodríguez, J. M. (2022). Desalination using the capacitive deionization technology with graphite/AC electrodes: Effect of the Flow Rate and Electrode Thickness. *Membranes*, *12*(7). https://doi.org/10.3390/membranes12070717

Nurohmah, A. R., Nisa, S. S., Stulasti, K. N. R., Yudha, C. S., Suci, W. G., Aliwarga, K., Widiyandari, H., & Purwanto, A. (2022). Sodium-ion battery from sea salt: a review. *Materials for Renewable and Sustainable Energy*, *11*(1), 71–89. https://doi.org/10.1007/s40243-022-00208-1

Panagopoulos, A. (2021). Water-energy nexus: desalination technologies and renewable energy sources. *Environmental Science and Pollution Research*, *28*(17), 21009–21022. https://doi.org/10.1007/s11356-021-13332-8

Pezeshki, Z., & Kettab, A. (2021). Desalination Battery. *Advances in Science, Technology and Innovation*, 137–162. https://doi.org/10.1007/978-3-030-72873-1_9

Porada, S., Borchardt, L., Oschatz, M., Bryjak, M., Atchison, J. S., Keesman, K. J., Kaskel, S., Biesheuvel, P. M., & Presser, V. (2013). Direct prediction of the desalination performance of porous carbon electrodes for capacitive deionization. *Energy and Environmental Science*, *6*(12), 3700–3712. https://doi.org/10.1039/c3ee42209g

Porada, S., Weinstein, L., Dash, R., Van Der Wal, A., Bryjak, M., Gogotsi, Y., & Biesheuvel, P. M. (2012). Water desalination using capacitive deionization with microporous carbon electrodes. *ACS Applied Materials and Interfaces*, *4*(3), 1194–1199. https://doi.org/10.1021/am201683j

Santos, C., & La Mantia, F. (2023). Insights into desalination battery concepts: current challenges and future perspectives. *Chemical Communications*, *59*(43), 6437–6452. https://doi.org/10.1039/d3cc00513e

Soffer, A., & Folman, M. (1972). The electrical double layer of high surface porous carbon electrode. *Journal of Electroanalytical Chemistry and Interfacial Electrochemistry*. https://www.sciencedirect.com/science/article/pii/S0022072872800871

Suss, M. E., Porada, S., Sun, X., Biesheuvel, P. M., Yoon, J., & Presser, V. (2015). Water desalination via capacitive deionization: What is it and what can we expect from it? *Energy & Environmental Science*, *8*(8), 2296–2319. https://doi.org/10.1039/C5EE00519A

Tang, W., Liang, J., He, D., Gong, J., Tang, L., Liu, Z., Wang, D., & Zeng, G. (2019). Various cell architectures of capacitive deionization: Recent advances and future trends. *Water Research*, *150*, 225–251. https://doi.org/10.1016/j.watres.2018.11.064

Wang, L., Zhang, Y., Moh, K., et al. (2021). From capacitive deionization to desalination batteries and desalination fuel cells. *Current Opinion in Electrochemistry*. https://www.sciencedirect.com/science/article/pii/S2451910321000727

Wang, L., Zhang, Y., Moh, K., & Presser, V. (2021). From capacitive deionization to desalination batteries and desalination fuel cells. *Current Opinion in Electrochemistry*, *29*(May), 100758. https://doi.org/10.1016/j.coelec.2021.100758

Wang, Z., Wang, Z., Lin, S., Jin, H., Gao, S., Zhu, Y., & Jin, J. (2018). Nanoparticle-templated nanofiltration membranes for ultrahigh performance desalination. *Nature.Com*. https://doi.org/10.1038/s41467-018-04467-3

Xing, W., Liang, J., Tang, W., He, D., Yan, M., Wang, X., Luo, Y., Tang, N., & Huang, M. (2020). Versatile applications of capacitive deionization (CDI)-based technologies. *Desalination*, *482*(January). https://doi.org/10.1016/j.desal.2020.114390

Xu, C., Yang, Z., Zhang, X., Xia, M., Yan, H., Li, J., et al. (2021a). Prussian blue analogues in aqueous batteries and desalination batteries. *Nano-Micro Letters*. https://link.springer.com/article/10.1007/s40820-021-00700-9

Xu, C., Yang, Z., Zhang, X., Xia, M., Yan, H., Li, J., Yu, H., Zhang, L., & Shu, J. (2021b). Prussian Blue Analogues in Aqueous Batteries and Desalination Batteries. In *Nano-Micro Letters* (Vol. *13*, Issue 1). Singapore: Springer. https://doi.org/10.1007/s40820-021-00700-9

Xu, D., Wang, W., Zhu, M., & Li, C. (2020). Recent advances in desalination battery: An Initial Review. *ACS Applied Materials and Interfaces*, *12*(52), 57671–57685. https://doi.org/10.1021/acsami.0c15413

Yu, F., Wang, L., Wang, Y., Shen, X., Cheng, Y., & Ma, J. (2019). Faradaic reactions in capacitive deionization for desalination and ion separation. *Journal of Materials Chemistry A*, *7*(27), 15999–16027. https://doi.org/10.1039/c9ta01264h

Yuan, L., Yang, X., Liang, P., Wang, L., Huang, Z. H., Wei, J., & Huang, X. (2012). Capacitive deionization coupled with microbial fuel cells to desalinate low-concentration salt water. *Bioresource Technology*, *110*, 735–738. https://doi.org/10.1016/j.biortech.2012.01.137

Zhang, C., He, D., Ma, J., Tang, W., & Waite, T. D. (2018). Faradaic reactions in capacitive deionization (CDI) - problems and possibilities: A review. *Water Research*, *128*(Cdi), 314–330. https://doi.org/10.1016/j.watres.2017.10.024

Zhang, X., Zhao, W., Zhang, Y., & Jegatheesan, V. (2021). A review of resource recovery from seawater desalination brine. *Reviews in Environmental Science and Biotechnology*, *20*(2), 333–361. https://doi.org/10.1007/S11157-021-09570-4

Desalination Plant Design and Energy Plant Optimization

8

8.1 INTRODUCTION

The global population is expected to expand by twenty-five% between 2015 and 2050, which will result in an even more rapid increase in water demand (F. Ahmed et al., 2020). The amount of freshwater necessary for human use is only a relatively small fraction of the total amount of water that exists on the earth, despite the fact that it is an abundant resource (Ayaz et al., 2022). In addition, the quantity of freshwater remains unchanged despite our water requirements increasing. It is estimated that millions of people lose their lives every year as a result of inadequate water supply and sanitation facilities. In order to assist in meeting the ever-increasing demand for water, numerous solutions have been proposed (Elimelech & Phillip, 2011). Some of these ideas include reusing more water, desalination, and better management of water resources. It is possible to desalinate brackish groundwater and seawater to create freshwater, which can help many communities suffering from drought worldwide (Aziz & Hanafiah, 2021).

Consequently, there is an increasing need for desalination on a global basis, as demonstrated by the rapid rise of the desalination sector over the previous few years. Desalination facilities worldwide generate approximately 95 m^3 of water daily at their peak. According to estimates, the nations located in the Middle East and North Africa (MENA) are responsible for 48% of the total desalination capacity constructed around the globe (Kajenthira and Murthy, 2013; Maftouh et al., 2023).

As a result of the reduction in membrane costs and the development of technology for energy recovery, reverse osmosis (RO) has emerged as the desalination method that is expanding at the most rapid rate (López-Ramírez et al., 2005). It is important to note that the amount of total dissolved solids (TDS) present in the feedwater considerably

DOI: 10.1201/9781003530893-8

influences the amount of energy that RO requires. The MED system is one of the first recognized systems for large-scale commercialized desalination (Ma et al., 2013). The process of MED involves the utilization of the latent heat that is produced by the steam in subsequent evaporators, which is also referred to as effects. This heat is then utilized to produce and condense vapor called product water. A water supply with high salinity is sprayed over the tubes utilized for heat transfer in the MED system (E. Ali et al., 2021). During the multi-stage flash (MSF) process, most of the liquid is heated to a boil, and the created pure vapor is subsequently condensed in the condenser tubes (El-Dessouky et al., 2000).

Consequently, MSF is less subject to scaling and deposit formation than MED (Lian-Ying et al., 2012). However, recent advancements in MED have rendered it more attractive than MSF. These developments include the introduction of scale inhibitory chemicals and the development of plate heat exchangers that are efficient. When it comes to satisfying the water demand, desalination is considered a relatively environmentally friendly, developed, and feasible technology (Darwish, 1991; Zahia et al., 2014).

When exploring suitable locations for desalination plants, one of the most fundamental concerns that needs to be addressed is the accessibility of an adequate supply of good-quality seawater (Huang et al., 2020). Compared to this, the availability of seawater to a desalination plant located along the coast is generally perceived as a comparatively straightforward factor to consider. It is usual practice to underestimate the relevance of an intake because of its impact on the location of a desalination plant, the overall layout of the plant, and the effectiveness with which it operates (Voutchkov, 2020).

Typically, the development and design of desalination plants begins with a thorough project planning study. Several environmental assessments and standard design steps are included in this study (Mittelman et al., 2007). These include the evaluation of the source water supply (including factors such as chemical composition, pH, salinity, as well as distance to the desalination facility), the possibilities for concentrate discharge, the pretreatment options, selecting the type of desalination method (including MSF, RO, MED, or ED), the sizing of the facility (including the determination of capacity), the estimation of power demands, the design of storage tanks (if necessary), and the distribution of water products (Druetta et al., 2013). Several engineering handbooks can provide comprehensive technical details regarding each design stage.

Most desalination plants are designed by selecting a specific technique and using heuristics and simulations to identify an optimal plant design. Membrane manufacturers offer engineering tools for designing RO plants using their membrane modules (Nafey, 1988). Utilizing specialized computer programs can facilitate a cost-effective assessment. The structural optimization method presents a promising alternative to the current state-of-the-art design process, particularly when considering the combination of several desalination processes (Qureshi & Zubair, 2016). Structural optimization enables the optimization of the operating point, major equipment parameters, and the procedure structure, defined by the presence of units and their relationship to one another (Skiborowski et al., 2012).

This chapter emphasizes the importance of selecting appropriate desalination technologies, like RO and multiple effect distillation (MED), and optimizing plant design to ensure efficiency and sustainability. The chapter also discusses the significance of intake and pretreatment processes, site selection, and the environmental impact of desalination plants. Additionally, it underscores the need for energy optimization, given that energy consumption is a significant cost factor in desalination. By integrating advanced technologies and optimizing operational parameters, desalination can become a more viable and environmentally friendly solution for providing freshwater to communities worldwide.

8.2 PLANT DESIGN

Designing a plant considers the project's physical, operating, and sustainability constraints (Reuters Staff, 2020). The process usually starts with coming up with several different project options. Then, the most viable option is chosen based on several factors, including the cost of capital and operating and maintenance, the impact on the surroundings, the carbon footprint of the building and its operation, the straightforwardness of operation and maintenance, the overall efficiency of the plant in terms of both energy and chemical consumption, the plant's ability to produce freshwater reliably, redundantly, and with spare capacity, its ability to grow and phased expansion, and its ability to adapt to new technologies and equipment (Abdelkareem et al., 2018).

8.2.1 Plant Capacity, Area, and Site

In the majority of cases, the freshwater production capacity of a project is determined by conducting a comprehensive comparison of the equilibrium between the water need in the working area of the production facility and the overall capacity of all available conventional sources of water supply and alternative sources of freshwater (such as desalination, rainwater harvesting, conservation, water importation, and reuse of water). In this way, the desalination plant can fulfill its water requirements during its planned period or usable life (Rivki et al., 2008).

Regarding water delivery, desalination is typically one of the more expensive options for a particular service region. The efficiency of a desalination plant is often assessed based on the volume of freshwater it can provide during prolonged scarcity, compared to other traditional and alternative water supply sources, as well as the additional costs associated with new water supplies (Voutchkov, 2020). Certain nations, including Israel, Australia, South Africa, Spain, and various countries in the region of the MENA, have established an internal objective of sourcing at least 25% to 50% of the total safe drinking water supply for their major coastal urban centers through seawater desalination (Andersen & Bank, 2012). This is an example of a country exposed to long shortage phases. Several of these nations already have brackish water desalination facilities that produce water from brackish surface water sources and saline inland aquifers that are near the maximum capacity they can produce (Khan et al., 2020).

When doing a capacity analysis for a desalination procedure, it is common practice to take into account the patterns of water supply on a daily and annual basis, the hydraulic restrictions of the water distribution system, the specific water quantity and quality requirements of indispensable individuals in the service region, and projections of future water demand (Fleming et al., 2005). Additionally, this investigation encompasses the specifications and expenses for transporting the water that has been desalinated to the distribution system, possible locations for connection and their corresponding capacity constraints, hydraulic system demands (such as pipe size and equipment size, as well as the operating pressure of the water distribution network at the desalinated water distribution stage), and constraints on the system's transportation capacity and possible solutions (Tan et al., 2022).

The selection of a location for a desalination plant is mainly determined by the availability of land that is located in close proximity to the principal users of desalinated water, as well as the location of transportation terminals for this water to the supply system (Fiorenza et al., 2007). It is frequently necessary to incorporate essential treatment equipment and systems to design a more efficient plant design. These equipment and systems include plant pretreatment filtering systems, energy recovery systems, RO racks, pumps, and chemical input and solids disposal systems in multistory buildings. The costs associated with this approach are often higher than those associated with installing all plant treatment facilities in single structures with slabs on the level (Gadhamshetty et al., 2014; Takabatake et al., 2021).

The following phase in planning a desalination project is to finish the preliminary plant design. After determining the operation area, location, and site of the desalination plant, as well as the quality of the water supply, the water products, and the concentrate, this stage is the next step (Huang et al., 2020). It has also been decided upon the configuration and the sort of intake and discharge that will be available. The preliminary design incorporates the basic plant site layout and hydraulic profile, the kind and order of treatment procedures and equipment, and the design criteria for the facility and equipment. The project implementation plan and estimates for the project's capital, operations, and maintenance costs are also included (Gude, 2018).

Within a traditional desalination plant, screens and filters are utilized to remove fine stones, dirt, and suspended or colloidal particles from the supply water. The removal of dissolved minerals, organic matter, and pathogens from the water comes next after this step in the process. When these two types of treatment processes, namely pretreatment and RO membrane separation, are utilized, the result is the formation of freshwater, often referred to as permeate, with a small amount of potentially harmful minerals and pathogens (K. P. Lee et al., 2011).

8.3 INTAKE TYPE AND SOURCE AND PRODUCT WATER QUALITY

Introducing salty water from a source is essential to every desalination plant. Throughout every phase of the desalination system's operational lifespan, the primary goal of the intake is to ensure that the source water is of an adequate and consistent supply and maintain its quality (Missimer & Maliva, 2018). For a desalination project, the choice of the intake type and configuration has a considerable impact on the concentration of foulants found in the water supply and the characteristics of those foulants. The complexity of the pretreatment process necessary to manage the fouling of the RO membrane is consequently affected. Surface or open intakes, underground intakes, and groundwater or well intakes are the most frequently utilized forms of desalination plant source water collecting facilities. Open intakes are pipes that connect an onshore or offshore input facility to a desalination plant (Missimer & Maliva, 2018). These pipelines collect source water straight from an open body, such as the ocean, a brackish river lake, or the sea (Fariñas & López, 2007). The source water obtained from these intakes often includes both large and small solid particles and silt and aquatic creatures such as fish, algae, and bacteria. These components contribute to the complexity of the water pretreatment process. The desalination plant receives water from a saltwater coastal aquifer or a brackish groundwater aquifer via underground intakes (Skiborowski et al., 2012). The aquifer's soils are used to transport source water as a consequence of subsurface intakes. The RO membrane's fouling potential is reduced by the natural removal of detritus, marine creatures, and some organics by these soils. As a result, pretreatment is typically not as involved when resource water is sourced from underground intakes rather than open ones (Bian et al., 2019).

Pretreatment is carried out with the purpose of reducing the number of suspended particles and silt that are present in the source water (Henthorne & Boysen, 2015). The specific values of this concentration may differ significantly between different projects. Certain desalination facilities, particularly those equipped with well intakes that collect water from uncontaminated saline aquifers unaffected by surface water pollution, may use minor pretreatment methods, such as cartridge or bag filtering. Rehardening and disinfection are the two types of water treatment processes included in the post-treatment of desalinated water. The RO filtrate undergoes re-hardening, increasing its hardness and bicarbonate alkalinity (Krista et al., 2009). This provides protection against corrosion for the distribution system that transports this water to the end consumers. The primary substances that introduce hardness and alkalinity to the permeate are calcium hydroxide (lime) and carbon dioxide, respectively. Recently, using limestone and carbon dioxide or sulfuric acid has become increasingly common as a post-treatment technology for corrosion prevention. This is because these technologies tend to generate water with lower turbidity (Greco et al., 2021).

The distance between the intake and the desalination operating site impacts the economic sustainability of the plant (Lyons et al., 2009). Additionally, the decisions about the pretreatment process are directly influenced by the quality and amount of available feedwater. Being the initial stage of the pretreatment process, the operation of an intake can have significant consequences on the whole functioning of the plant and its influence on the marine environment. Accepting a new desalination plant could rely on the methods used to solve environmental problems connected to seawater intake (Rodríguez-Girones et al., 1996). Furthermore, the prospective loss of recreational activities in the intake area and the visual effects of particular intake layouts are considered. An optimal intake design not only safeguards downstream equipment and minimizes the ecological consequences on marine life but also enhances the efficiency and lowers the operational expenses of the pretreatment systems (Mohammed & Askalany, 2019).

8.3.1 Surface Water Intakes

In the past, large-scale seawater desalination plants would use open-ocean surface water intakes that appeared very similar to the screens used to collect condenser cooling water by electric power plants (Azmi et al., 2024). These screens were mechanically cleaned. A screening chamber and pump station are usually located on land and connected to the open ocean via a long input pipe, a concrete canal, or a pier (Missimer et al., 2018).

8.3.2 Traveling Water Screens

In the 1890s, the first seawater intakes had water screens that could move between locations. Each screen is outfitted with wire mesh panels that rotate and apertures ranging from 6 mL to 9.5 mL. The debris that has been gathered is removed by a high-pressure water spray, which is then rinsed into a trough for further disposal (Huang et al., 2020). This is accomplished by rotating the wire mesh panels out of the flow. Onshore, at the end of a channel or forebay that extends beyond the surf zone, or at the end of a pipe that continues into the sea and winds up with a vertical entrance known as a "velocity cap," are all possible locations for the installation of screens (Ludwig, 2022).

8.3.2.1 Passive Screens

Another intake configuration includes slotted screens composed of trapezoidal-shaped "wedge wire." Cylindrical screens having openings from 0.5 mL to 10 mL are generally oriented horizontally and designed to maintain a velocity below 15 cm/s (0.5 ft/s) to reduce the incursion of debris and marine organisms (Azmi et al., 2024; Missimer et al., 2015). Passive screens are most appropriate for locations with an existing ambient cross-flow current. When debris buildup occurs, an air backwash system is generally advised to remove the screens. When selecting materials for submerged equipment, it is essential to consider the possibilities for corrosion and biofouling that occur in seawater (Missimer et al., 2015). This is true for all equipment that is submerged. Empirical research has demonstrated that passive screens are quite good at reducing impingement and entrainment (Voutchkov, 2020). The combination of their slot width and the low velocity required for through-flow is directly related to their effectiveness. Experimental evidence has shown that apertures measuring 1 mm are quite efficient in excluding larvae and reducing entrainment by more than 80% (Huang et al., 2020).

8.3.2.2 Fine Mesh Traveling Screens

Fine mesh traveling screens have efficiently reduced the entrainment of eggs, larvae, and juvenile fish at specific intake locations. This has been accomplished by fitting traveling water screens with wire mesh panels with openings ranging from 0.5 mL to 5 mL (Federation) & 1980, 1980). There is a possibility that this adjustment could reduce entrainment by as much as 80%. The use of fine mesh screens, on the other hand, might generate operational difficulties since it makes removing debris more difficult, which in turn has an impact on marine life. Additionally, in certain areas, the fine mesh is only employed on a seasonal basis, specifically during high egg and larval growth (Walraven & Jager, 2020).

8.3.3 Subsurface Intakes

Other subsurface intakes include infiltration galleries, seabed filtration systems, and beach wells, which can be either horizontal or vertical in orientation. The open ocean is separated from the site of intake by a geologic unit, a component of each notion (Bourouni et al., 2011). Because the geological conditions under surface water are usually impermeable or of sufficient thickness and depth to permit water extraction, a subsurface intake may be utilized in certain circumstances. Additionally, apart from offering inherent filtration, this configuration offers the benefit of effectively isolating the majority of marine creatures from the water input (Carravetta et al., 2016). Subterranean intakes can be assessed and controlled as groundwater sources in certain instances. Implementing subsurface intakes provides a clear environmental benefit, effectively eliminating the biological consequences of marine life's impingement and entrainment. Yet, subsurface designs must consider their possible adverse effects on adjacent freshwater aquifers (Rivki et al., 2008; Voutchkov, 2020).

8.3.3.1 Radial Well

A modified form of the beach well technique, radial collector wells involve the connection of multiple horizontal collector wells to a central caisson. In addition to its function as a pumping station or wet well, this caisson is responsible for transporting water to the desalination plant. Using many horizontal wells, each radial well can produce a significantly larger production than a single vertical well. A direct-jack or pull-back drilling technique might be utilized to drill individual horizontal wells or hydraulically raise well screens out from the bottom of the caisson. With radial arms that commonly measure between 200 mL and 300 mL (8–12 in.) in diameter and a depth that ranges from 9 m to 45 m (30–150 ft), caissons can have a diameter that ranges from 2.75 m to 6 m (9–20 ft) (Hansen & Nath, 1978; Houben et al., 2022).

8.3.3.2 Vertical Beach Wells

One type of shallow intake well is the vertical beach well, which uses geological formations or beach sand as its filter media. For desalination plants with flow rates lower than 20,000 m^3/d, beach wells can replace open sea intakes at a lower cost. Their benefit lies in providing "pre-filtered" water, which can significantly decrease the need for further pretreatment protocols. In addition to a well screen and a vertical turbine pump, a vertical beach well is constructed with a casing that is built of materials that are not metallic (Houben et al., 2022). To assess site appropriateness, test wells are drilled, and a thorough hydrogeologic study is carried out to establish the formation's transmissivity and the substrate's features. For optimal results, beach wells should be positioned in close proximity to the coastline. The maximum yield from each well varies between 0.1 m^3 and 4000 m^3/d (0.1–1.0 mgd) (Prathapaneni & Detroja, 2020).

8.4 WATER QUALITY

The capacity for desalinating seawater differs depending on the specific desalination method used. The volume needed for feedwater may vary from around double the production capacity of RO systems to over ten times the amount of distillate generated by thermal processes, which typically have requirements for both process and cooling water (Uziel, 1980). Continuous availability of the required feedwater is essential for a plant to achieve its productivity objectives while it is in operation (Sattarkhan et al., 2024).

Most saltwater desalination facilities have utilized MSF evaporation or MED desalination technologies. These facilities have also been located in close proximity to an electric power-producing facility that shares a waterway or nearshore seawater intake. As a result of the fact that the condenser tubes used in a power plant and those used in a medical facility or a medical supply facility are of comparable size, both types of systems require just a small amount of treatment (Uche et al., 2001). Utilizing a moving water screen or a rotating drum screen with wire mesh apertures measuring between 6 mL and 9.5 mL is typically the most effective method. When compared to this, the efficiency of a seawater reverse osmosis (SWRO) plant can be significantly improved by ensuring that the water quality is more consistent and employing a more exact screening method (Vince et al., 2008).

There is a greater emphasis on selecting a design or location that would give a consistent saltwater quality and limit the presence of suspended solids. Most SWRO plants are independent operations with a specifically built input (Helal et al., 2003). The intake efficiency is the first stage in the SWRO pretreatment process, and it has the potential to significantly impact the overall operation and performance of the plant during this stage. Desalination facilities demonstrate optimal efficiency and predictability when the properties of the feedwater are reasonably stable and are not prone to sudden or significant changes in water quality, similar to other process systems. As a result, water quality evaluation must consider daily and seasonal fluctuations (Water, 2019). An exhaustive analysis of past information on water quality, focusing primarily on the total suspended solids (TSS), ocean water temperature, TDS, and total organic carbon (TOC), should be included as an essential component of the evaluation (Cotruvo, 2018; Omer, 1975). This analysis should be performed to identify all of the factors that can potentially influence the workings of the plant and the functionality of the process. Most SWRO) facilities convert 40–50% of the water brought into product water. The remaining water containing the salt removed by the RO system is then returned to the sea for controlled release. Because of this, it is beneficial to locate the desalination facility close to the beach to limit the amount of pumping required for intake and release (Alam et al., 2007).

The quality of product water is a crucial determinant that significantly influences the design and finance of a plant. The following analysis of significant product water quality concerns must be considered during the brackish and seawater desalination system design (Gupta et al., 2003). Depending on factors such as water supply temperature and ion composition, the layout of the RO membrane approach, and the salt rejection capabilities of the membranes used for desalination, the mineral content of the permeate that SWRO desalination facilities produce can vary significantly. This is known as the mineral content of the permeate. The quality of the permeate water produced by SWRO systems with different configurations (single-pass, entire two-pass, and split permeate second pass) from various seawater sources was analyzed (Batabyal & Chakraborty, 2015).

Desalinated saltwater produced by SWRO systems often has substantially lower organic content than most conventional freshwater sources, such as rivers, lakes, and groundwater aquifers (Dohare et al., 2014). This is because SWRO systems can remove organic matter from seawater. Consequently, the concentration of disinfection byproducts (DBPs) in desalinated water decreases significantly when chlorine is used to disinfect the water. One potential threat to the quality of desalinated water generated from surface water sources is the presence of reduced molecular weight algal toxins. Some examples of these toxins include domoic acid and saxitoxin (Yermiyahu et al., 2007). The creation of these toxins occurs during algal blooms when the concentration of algae in the saltwater can multiply by a factor of several hundred. Domoic acid and saxitoxin are essential among the various organic toxins produced during algal blooms. These toxins include okadaic acid, brevetoxin, yessotoxin, microcystin, and nodularin. This is because their molecular size and weight are comparable in relation to the average molecule rejection size (molecular weight cutoff) of SWRO membranes. In addition, it is conceivable that some of these toxins could potentially get through the membranes under certain circumstances (Damodhar & Vikram Reddy, 2013).

The utilization of RO desalination systems in combination with various pre- and post-RO water treatment procedures currently in use in the commercial sector makes it possible to generate water of nearly any specified grade (Kress et al., 2020). In general, the optimal water quality for a desalination plant is determined by considering the regulations that govern the water produced as a final product and the water quality needs of the major water consumers in the region that the facility serves. Suppose the requirements are largely industrial, agricultural, or horticultural in nature. In that case, the required quality of the water produced through desalination may, in certain circumstances, be higher than that of drinking water. In most cases, desalination plants used for municipal purposes are intended to produce water that satisfies the minimal requirements for drinking water (Dohare et al., 2014). This is especially true when the water is the primary supply source for the entire service area. Despite this, countless utilities and municipalities worldwide use desalinated water as an extra source of water supply. This water is then blended with other municipal water sources already in the distribution system (Verma & Singh, 2013).

Given the relatively simple process of adjusting the quality of desalinated water to a desired level compared to other conventional water sources, it is common practice to produce desalinated water at a higher quality than the other sources and then blend it to enhance the final product water supplied to consumers. Although desalinated water production incurs higher costs, this method is often the most economically efficient overall strategy for enhancing water quality throughout the distribution system, compared to implementing supplementary treatment procedures at individual traditional water treatment plants working in the same service area (Cotruvo, 2018).

Effective management of concentrate and other waste streams related to desalinated water production is a crucial project design element influencing the plant's geographical placement, scale, and treatment procedures. Typically, a desalination plant site must be situated within an acceptable distance (averaging 0.5–10 km or 0.3–6.1 m) of appropriate locations for concentrate disposal to be viable. To be considered appropriate, a plant discharge disposal site must possess a physical layout and receiving capacity that enables the continuous environmentally safe disposal of the concentrate and, if feasible, other plant waste streams throughout the whole operational lifespan of the desalination project (Wilf & Bartels, 2005).

8.5 ENERGY PLANT OPTIMIZATION

The primary determinant for the successful implementation of any technology in desalination is energy consumption, which accounts for over 50–60% of the total expenditures (S. Lee et al., 2018). Presently, desalination facilities produce 5–26 times the amount of work estimated as the theoretical minimum. For this reason, it is necessary to improve both the technology and economies of scale to enhance the energy efficiency of desalination operations. The amount of energy that is required for desalination operations is affected by several different parameters. The capacity and design of the plant, the quality of the feed saltwater, and the materials that are used are all included in these factors (Onishi & Mujtaba, 2021). It is of the utmost importance to emphasize that the quantity of energy needed in thermal-based desalination procedures, such as MSF, is not affected by the concentration of salt in the saline used as feed. On the other hand, energy needs are much higher in membrane-based desalination processes, like RO, because of the increased osmotic pressure that occurs when the feed salinity is higher (Ghobeity & Mitsos, 2014). The principal energy sources utilized by RO systems are electricity obtained from fossil fuels (thermal coal, natural gas, and petroleum), nuclear power, and energy from renewable sources (Goh et al., 2021).

On the other hand, MSF technologies rely on thermal energy as their primary energy source, which is produced from waste heat from power plants, heat sources from industrial waste, boilers powered by fossil fuels, and sustainable energy sources. Electricity is utilized as a secondary source to operate pumps associated with the MSF system. Existing literature indicates that the MSF process demands more energy than the RO process. The MSF method, in particular, calls for 19.58–27.25 kWhe/m^3 of water, but the SWRO process, which incorporates an energy recovery system, calls for 4–6 kWhe/m^3 of water (F. E. Ahmed et al., 2019; Goh et al., 2021).

Using evolutionary algorithms to solve multi-objective optimization problems, Ali and Kairouani (2015) modified the operational parameters of a brine recirculation MSF plant. They did this by considering changes in the fouling factor of the brine heater as well as seasonal variations in the temperature of the seawater. Steam temperature (T_s), mass flow rate (M_R), mass flow rate of cooling water (M_{CW}), and mass flow rate of feed (M_f) were the optimal operational parameters for the boiler steam, recycled brine, cooling seawater, and make-up seawater flows, respectively. A nominal 26,700 m^3/day output capacity is provided by the used plant data, which features 16 flashing stages. To achieve these objectives, we optimized the capacity to produce freshwater, reduced the flow rate of heating steam to manage thermal energy consumption, and minimized the flow rates of pumps to limit electrical energy consumption. The best possible operational parameter combinations were used to build an ensemble of Pareto optimum solutions. A steady-state process model of the plant was used to solve the optimization problem. This model included heat transfer equations, mass and energy balances, and other relevant parameters.

8.6 CONCLUSION

The chapter on desalination plant design and energy optimization highlights the critical role of desalination in addressing the global water scarcity challenge. With the world's population projected to grow significantly by 2050, the demand for freshwater is expected to rise sharply. Desalination, improved water resource management, and increased water reuse offer a viable solution to meet this demand. The chapter discusses various desalination technologies, including RO, multi-effect distillation (MED), and MSF, emphasizing the advancements in energy recovery and membrane technology that have made RO the fastest-growing method. The design of desalination plants involves carefully considering several factors, such as the availability and quality of seawater, environmental impact, cost, and energy efficiency. The chapter also explores optimizing energy consumption, as it constitutes a significant portion of the operational costs. By employing advanced optimization techniques and leveraging renewable energy sources, the efficiency of desalination processes can be enhanced, making them more sustainable and cost-effective. The chapter underscores the importance of desalination as a critical component in the global strategy to ensure a reliable and sustainable water supply for future generations.

REFERENCES

Abdelkareem, M. A., El Haj Assad, M., Sayed, E. T., & Soudan, B. (2018). Recent progress in the use of renewable energy sources to power water desalination plants. *Desalination*, *435* (November), 97–113. https://doi.org/10.1016/j.desal.2017.11.018

Ahmed, F. E., Hashaikeh, R., Diabat, A., & Hilal, N. (2019). Mathematical and optimization modelling in desalination: State-of-the-art and future direction. *Desalination, 469*. https://doi.org/10.1016/j.desal.2019.114092

Ahmed, F., Hashaikeh, R., & Hilal, N. (2020). Hybrid technologies: The future of energy efficient desalination–A review. *Desalination*. https://www.sciencedirect.com/science/article/pii/S0011916420313370

Alam, M. J. B., Islam, M. R., Muyen, Z., Mamun, M., & Islam, S. (2007). Water quality parameters along rivers. *International Journal of Environmental Science and Technology, 4*(1), 159–167. https://doi.org/10.1007/BF03325974

Ali, E., Orfi, J., AlAnsary, H., Soukane, S., Elcik, H., Alpatova, A., & Ghaffour, N. (2021). Cost analysis of multiple effect evaporation and membrane distillation hybrid desalination system. *Desalination, 517*(July), 115258. https://doi.org/10.1016/j.desal.2021.115258

Ali, M., & Kairouani, L. (2015). Multi-objective optimization of operating parameters of a MSF-BR desalination plant using solver optimization tool of Matlab software. *Desalination*. https://doi.org/10.1016/j.desal.2015.11.030

Andersen, I., & Bank, W. (2012). Environmental Impacts of Desalination. *Renewable Energy Desalination: An Emerging Solution To Close the Water Gap in the Middle East and North Africa*, 111–126.

Ayaz, M., Namazi, M., Din, M. et al. (2022). Sustainable seawater desalination: Current status, environmental implications and future expectations. *Desalination*. https://www.sciencedirect.com/science/article/pii/S0011916422004775

Aziz, N. I. H. A., & Hanafiah, M. M. (2021). Application of life cycle assessment for desalination: Progress, challenges and future directions. *Environmental Pollution, 268*, 115948. https://doi.org/10.1016/j.envpol.2020.115948

Azmi, A. A., Yasunaga, T., Fontaine, K., Morisaki, T., Nakaoka, T., Thirugnana, S. T., Jaafar, A. B., & Ikegami, Y. (2024). Basic design optimization of power and desalinated water for hybrid cycle ocean thermal energy conversion system integrated with desalination plant. *Journal of Marine Science and Technology (Japan), 29*(2), 333–352. https://doi.org/10.1007/s00773-024-00988-3

Batabyal, A., & Chakraborty, S. (2015). Hydrogeochemistry and water quality index in the assessment of groundwater quality for drinking uses. *Wiley Online Library, 87*(7), 607–617. https://doi.org/10.2175/106143015X14212658613956

Bian, D. W., Watson, S. M., Wright, N. C., Shah, S. R., Buonassisi, T., Ramanujan, D., Peters, I. M., & Winter, A. G. (2019). Optimization and design of a low-cost, village-scale, photovoltaic-powered, electrodialysis reversal desalination system for rural India. *Desalination, 452*(August 2018), 265–278. https://doi.org/10.1016/j.desal.2018.09.004

Bourouni, K., Ben M'Barek, T., & Al Taee, A. (2011). Design and optimization of desalination reverse osmosis plants driven by renewable energies using genetic algorithms. *Renewable Energy, 36*(3), 936–950. https://doi.org/10.1016/j.renene.2010.08.039

Carravetta, A., Fecarotta, O., Golia, U. M., La Rocca, M., Martino, R., Padulano, R., & Tucciarelli, T. (2016). Optimization of Osmotic Desalination Plants for Water Supply Networks. *Water Resources Management, 30*(11), 3965–3978. https://doi.org/10.1007/s11269-016-1404-9

Cotruvo, J. (2018). *Drinking Water Quality and Contaminants Guidebook*. CRC Press. https://doi.org/10.1201/9781351110471/DRINKING-WATER-QUALITY-CONTAMINANTS-GUIDEBOOK-JOSEPH-COTRUVO

Damodhar, U., & Vikram Reddy, M. (2013). Impact of pharmaceutical industry treated effluents on the water quality of river Uppanar, South east coast of India: A case study. *Applied Water Science, 3*(2), 501–514. https://doi.org/10.1007/s13201-013-0098-x

Darwish, M. A. (1991). Thermal analysis of multi-stage flash desalting systems. *Desalination*, *85*(1), 59–79. https://doi.org/10.1016/0011-9164(91)85147-M

Dohare, D., Deshpande, S., & Kotiya, A. (2014). Analysis of ground water quality parameters: A review. *Research Journal of Engineering Sciences*, *3*(5), 2278–9472. http://www.isca.me/IJES/Archive/v3/i5/3.ISCA-RJEngS-2014-24.pdf

Druetta, P., Aguirre, P., & Mussati, S. (2013). Optimization of Multi-Effect Evaporation desalination plants. *Desalination*, *311*, 1–15. https://doi.org/10.1016/j.desal.2012.10.033

El-Dessouky, H. T., Ettouney, H. M., & Mandani, F. (2000). Performance of parallel feed multiple effect evaporation system for seawater desalination. *Applied Thermal Engineering*, *20*(17), 1679–1706. https://doi.org/10.1016/S1359-4311(99)00098-8

Elimelech, M., & Phillip, W. A. (2011). The future of seawater desalination: Energy, technology, and the environment. *Science*, *333*(6043), 712–717. https://doi.org/10.1126/science.1200488

Fariñas, M., & López, L. Á. (2007). New and innovative sea water intake system for the desalination plant at San Pedro del Pinatar. *Desalination*, *203*(1–3), 199–217. https://doi.org/10.1016/j.desal.2006.01.036

Uziel, M. S. (1980). Entrairtment and impingement at cooling water intakes. *Journal (Water Pollution Control Federation)*. https://www.jstor.org/stable/40586479

Fiorenza, G., Sharma, V. K., & Braccio, G. (2007). Techno-economic evaluation of a solar powered water desalination plant. *NATO Security through Science Series C: Environmental Security*, 33–41. https://doi.org/10.1007/978-1-4020-5508-9_3

Fleming, K. K., Gullick, R. W., Dugandzic, J. P., & Lechevallier, M. W. (2005). *Susceptibility of Potable Water Distribution Systems to Negative Pressure Transients*. https://rucore.libraries.rutgers.edu/rutgers-lib/35401/

Gadhamshetty, V., Gude, V. G., & Nirmalakhandan, N. (2014). Thermal energy storage system for energy conservation and water desalination in power plants. *Energy*, *66*, 938–949. https://doi.org/10.1016/j.energy.2014.01.046

Ghobeity, A., & Mitsos, A. (2014). Optimal design and operation of desalination systems: New challenges and recent advances. *Current Opinion in Chemical Engineering*, *6*, 61–68. https://doi.org/10.1016/j.coche.2014.09.008

Goh, P. S., Liang, Y. Y., & Ismail, A. F. (2021). Energy Efficient Seawater Desalination: Strategies and Opportunities. *Energy Technology*, *9*(5), 1–26. https://doi.org/10.1002/ente.202100008

Greco, F., Heijman, S. G. J., & Jarquin-Laguna, A. (2021). Integration of wind energy and desalination systems: A review study. *Processes*, *9*(12), 1–29. https://doi.org/10.3390/pr9122181

Gupta, A., Gupta, S., & Patil, R. (2003). A comparison of water quality indices for coastal water. *Journal of Environmental Science and Health, Part A*. https://doi.org/10.1081/ESE-120024458

Hansen, R. M., & Nath, J. H. (1978). *Optimizing intake screens for Ocean Thermal Energy Conversion power plants*. https://ir.library.oregonstate.edu/downloads/sb397c286

Helal, A. M., El-Nashar, A. M., Al-Katheeri, E., & Al-Malek, S. (2003). Optimal design of hybrid RO/MSF desalination plants part I: Modeling and algorithms. *Desalination*, *154*(1), 43–66. https://doi.org/10.1016/S0011-9164(03)00207-8

Henthorne, L., & Boysen, B. (2015). State-of-the-art of reverse osmosis desalination pretreatment. *Desalination*, *356*, 129–139. https://doi.org/10.1016/j.desal.2014.10.039

Houben, G. J., Collins, S., Bakker, M., Daffner, T., Triller, F., & Kacimov, A. (2022). Review: Horizontal, directionally drilled and radial collector wells. *Hydrogeology Journal*, *30*(2), 329–357. https://doi.org/10.1007/S10040-021-02425-W

Huang, B., Pu, K., Wu, P., Wu, D., & Leng, J. (2020). Design, selection and application of energy recovery device in seawater desalination: a review. *Energies*, *13*(6). https://doi.org/10.3390/en13164150

Kajenthira, A., & Murthy, S. (2013). Urban water challenges in the MENA region: integrating Islamic principles with demand management strategies. *ElgarOnline*. https://www.elgaronline.com/downloadpdf/edcoll/9781781006412/9781781006412.00011.pdf

Khan, Q., Maraqa, M. A., & Mohamed, A. M. O. (2020). Inland desalination. In *Pollution Assessment for Sustainable Practices in Applied Sciences and Engineering*. https://doi.org/10.1016/B978-0-12-809582-9.00017-7

Kress, N., Gertner, Y., & Shoham-Frider, E. (2020). Seawater quality at the brine discharge site from two mega size seawater reverse osmosis desalination plants in Israel (Eastern Mediterranean). *Water Research*, *171*. https://doi.org/10.1016/j.watres.2019.115402

Krista, L., Kulinkina, A., & Plummer, J. D. (2009). *The Design of a Desalination Pretreatment System For Brackish Groundwater A Major Qualifying Project Report.*

Lee, K. P., Arnot, T. C., & Mattia, D. (2011). A review of reverse osmosis membrane materials for desalination-Development to date and future potential. In *Journal of Membrane Science* (Vol. 370, Issues 1–2, pp. 1–22). Elsevier. https://doi.org/10.1016/j.memsci.2010.12.036

Lee, S., Cho, H. Y., & Har, D. (2018). Operation optimization with jointly controlled modules powered by hybrid energy source: A case study of desalination. *Renewable and Sustainable Energy Reviews*, *81*(August), 3070–3080. https://doi.org/10.1016/j.rser.2017.08.076

Lian-Ying, W., Sheng-Nan, X., & Cong-Jie, G. (2012). Simulation of Multi-stage Flash(MSF) Desalination Process. *Advances in Materials Physics and Chemistry*, *2*(04), 200–205. https://doi.org/10.4236/ampc.2012.24b052

López-Ramírez, J. A., María Quiroga, J., López-Ramírez, J. A., Oviedo, C., & Quiroga Alonso, J. (2005). Comparative studies of reverse osmosis membranes for wastewater reclamation. *Elsevier*, *191*, 137–147. https://doi.org/10.1016/j.desal.2005.08.013

Ludwig, H. (2022). Reverse Osmosis Seawater Desalination Volume 1: Planning, Process Design and Engineering - A Manual for Study and Practice. In *Reverse Osmosis Seawater Desalination Volume 1: Planning, Process Design and Engineering - A Manual for Study and Practice* (Vol. *1*). https://doi.org/10.1007/978-3-030-81931-6

Lyons, E., Zhang, P., Benn, T., Sharif, F., Li, K., Crittenden, J., Costanza, M., & Chen, Y. S. (2009). Life cycle assessment of three water supply systems: Importation, reclamation and desalination. *Water Science and Technology: Water Supply*, *9*(4), 439–448. https://doi.org/10.2166/ws.2009.534

Ma, Z. Q., Huo, S. H., & Su, M. (2013). Simulation with Aspen Plus and performance analysis of LT-MED seawater desalination system. *Applied Mechanics and Materials*, *397–400*, 948–956. https://doi.org/10.4028/www.scientific.net/AMM.397-400.948

Maftouh, A., Fatni, O. El, Bouzekri, S., Bahaj, T., Kacimi, I., Hajjaji, S. El, & Malik, A. (2023). Solar Desalination: Current Applications and Future Potential in MENA Region – A Case Study. *Journal of Sustainable Development of Energy, Water and Environment Systems*, *11*(2), 1–26. https://doi.org/10.13044/j.sdewes.d10.0437

Missimer, T. M., Hogan, T. W., & Pankratz, T. (2015). Passive screen intakes: Design, construction, operation, and environmental impacts. *Environmental Science and Engineering (Subseries: Environmental Science)*, *149*, 79–104. https://doi.org/10.1007/978-3-319-13203-7_5

Missimer, T. M., & Maliva, R. G. (2018). Environmental issues in seawater reverse osmosis desalination: Intakes and outfalls. *Desalination*, *434*(April), 198–215. https://doi.org/10.1016/j.desal.2017.07.012

Missimer, T. M., Watson, I., Maliva, R. G., Ghaffour, N., Dehwah, A. H. A., Woolschlager, J., & Hegy, M. (2018). Impacts of natural pore-water and offshore aquifer chemistry on the operation and economics of some subsurface intakes types for SWRO plants. *Desalination and Water Treatment*, *132*, 1–9. https://doi.org/10.5004/dwt.2018.23134

Mittelman, G., Mouchtar, O., & Dayan, A. (2007). Large-scale solar thermal desalination plants: A review. *Heat Transfer Engineering*, *28*(11), 924–930. https://doi.org/10.1080/01457630701421711

Mohammed, R. H., & Askalany, A. A. (2019). Productivity improvements of adsorption desalination systems. In *Green Energy and Technology* (pp. 325–357). https://doi.org/10.1007/978-981-13-6887-5_15

Nafey, A. S. M. T. (1988). *Design and simulation of seawater thermal desalination plants. November.* https://etheses.whiterose.ac.uk/15208/

Omer, N. H. (1975). Water Quality Parameters. *ASTM Special Technical Publication, STP 573*, 1–18. https://doi.org/10.1201/9781439800539.axb

Onishi, V. C., & Mujtaba, I. M. (2021). Special issue "design, control and optimization of desalination processes". *Processes*, *9*(3), 11–12. https://doi.org/10.3390/pr9030541

Prathapaneni, D. R., & Detroja, K. (2020). Optimal design of energy sources and reverse osmosis desalination plant with demand side management for cost-effective freshwater production. *Desalination*, *496*(May), 114741. https://doi.org/10.1016/j.desal.2020.114741

Qureshi, B. A., & Zubair, S. M. (2016). Exergetic efficiency of NF, RO and EDR desalination plants. *Desalination*, *378*, 92–99. https://doi.org/10.1016/j.desal.2015.09.027

Reuters Staff. (2020). *Inside the world's biggest water desalination plants | Reuters*. Reuters. https://www.reuters.com/article/us-saudi-water-desalination-idUSKBN26Y1HD

Rivki, M., Bachtiar, A. M., Informatika, T., Teknik, F., & Indonesia, U. K. (2008). Desalination Engineer. In *Journal AWWA* (Vol. 100, Issue 1). https://doi.org/10.1002/j.1551-8833.2008.tb08114.x

Rodríguez-Girones, M., Pérez, J., & Veza, J. (1996). A systematic approach to desalination powered by solar, wind and geothermal energy sources. *In*: *Proceedings of the Mediterranean Conference on Renewable Energy Sources for Water Production. European Commission, EURORED Network, CRES, EDS*, 20–25. https://scholar.google.com/scholar?lookup=0&q=Rodriguez-Girones,+Rodriguez+M,+Perez+J,+Veza+J.+A+systematic+approach+to+desalination+powered+by+solar,+wind+and+geothermal+energy+sources&hl=en&as_sdt=0,5#0

Sattarkhan, M. H., Mostafaeipour, A., & Sadegheih, A. (2024). A novel mathematical model for simultaneous optimization of desalination plant location and water distribution network; A case study. *Heliyon*, *10*(12), e32758. https://doi.org/10.1016/j.heliyon.2024.e32758

Skiborowski, M., Mhamdi, A., Kraemer, K., & Marquardt, W. (2012). Model-based structural optimization of seawater desalination plants. *Desalination*, *292*, 30–44. https://doi.org/10.1016/j.desal.2012.02.007

Gude, V. G. (2018). *Sustainable Desalination Handbook: Plant Selection, Design and Implementation*. https://doi.org/10.1016/C2015-0-04758-X

Takabatake, H., Taniguchi, M., & Kurihara, M. (2021). Advanced technologies for stabilization and high performance of seawater RO membrane desalination plants. *Membranes*, *11*(2), 1–20. https://doi.org/10.3390/membranes11020138

Tan, Y., Chen, H., & He, R. (2022). Water distribution and transport-kinetics model in fresh cement-based mixtures containing superabsorbent polymers based on 1H low-field NMR. *Cement and Concrete Composites*, *127*(10439), 6. https://doi.org/10.1016/j.cemconcomp.2021.104396

Uche, J., Serra, L., & Valero, A. (2001). Thermoeconomic optimization of a dual-purpose power and desalination plant. *Desalination*, *136*(1–3), 147–158. https://doi.org/10.1016/S0011-9164(01)00177-1

Verma, A. K., & Singh, T. N. (2013). Prediction of water quality from simple field parameters. *Environmental Earth Sciences*, *69*(3), 821–829. https://doi.org/10.1007/S12665-012-1967-6

Vince, F., Marechal, F., Aoustin, E., & Bréant, P. (2008). Multi-objective optimization of RO desalination plants. *Desalination*, *222*(1–3), 96–118. https://doi.org/10.1016/j.desal.2007.02.064

Voutchkov, N. (2020). Introduction to Desalination Project Design and Delivery. *Sun Cam.*

Walraven, L. van, & Jager, Z. (2020). *Review of the possible interactions between Blue Energy and aquatic*. https://doi.org/10.25850/nioz/7b.b.2

Water, N. (2019). *Drinking water quality*. https://niopa.qub.ac.uk/bitstream/NIOPA/9434/4/2017 niwaterdrinkingwaterqualityannualreport.pdf

Wilf, M., & Bartels, C. (2005). Optimization of seawater RO systems design. *Desalination*, *173*(1), 1–12. https://doi.org/10.1016/j.desal.2004.06.206

Yermiyahu, U., Tal, A., Ben-Gal, A., Bar-Tal, A., Tarchitzky, J., & Lahav, O. (2007). Environmental science: Rethinking desalinated water quality and agriculture. *Science*, *318*(5852), 920–921. https://doi.org/10.1126/science.1146339

Zahia, T., Hanane, A., Randha, B., Djamila, Z., & Mohamed, A. (2014). Water desalination system: Experimental study on the performance of a multi-stage still. *Applied Mechanics and Materials*, *548–549*, 546–549. https://doi.org/10.4028/www.scientific.net/AMM.548-549.546

Overview of the Health and Environmental Impacts of Desalination

9

9.1 INTRODUCTION

The growing global population and rapid urbanization, industrialization, and progress have resulted in over-exploiting natural resources (Miller et al., 2015). Furthermore, the exponential expansion of socioeconomic activity might lead to environmental degradation, including the contamination of water sources (Zhou et al., 2011). During the early stages of industrialization, the global population was approximately seven hundred million (Andersen & Bank, 2012). The current world population exceeds seven billion and is anticipated to exceed nine billion by 2050, reaching at least ten billion by the beginning of the twenty-second century. A significant portion of the population increase is expected to occur in Africa and Asia, where there is already an increasing need for potable water for farming, manufacturing, and other development-related operations across all sectors (Sirota et al., 2024).

Around 97% of the water on Earth is salty, which means it cannot be used for drinking or farming. The global population growth and rise in freshwater demand are essential factors contributing to the decline and limited availability of freshwater resources (Lattemann & Höpner, 2008). Recent technological developments have led to the recognition of saltwater in seas and oceans as a crucial resource for satisfying the present need for freshwater. Various approaches and technologies have been suggested to address global water challenges (Prajapati et al., 2021). These include water reuse and recycling, use of advanced wastewater processing technologies, rainwater harvesting, collecting and preserving rainwater for future use, use of effective irrigation techniques

DOI: 10.1201/9781003530893-9

such as drip irrigation and precision agriculture, implementation of innovative water management technologies like sensors, data analytics, and remote monitoring, promotion of water conservation through training, and desalination plants (Ghobeity & Mitsos, 2014; Lyons et al., 2009). Among these approaches, desalination is one of the most significant and effective methods. Worldwide, 16,436 desalination facilities provide more than 92.5 million m^3 of freshwater to nearby communities daily, according to the International Desalination Association (IDA) (Voutchkov et al., 2019).

While desalination methods show great potential and have been widely adopted as a primary alternative for countries with limited water resources in coastal areas, they do have inevitable environmental consequences (Koroneos et al., 2007). There is a growing worldwide interest in understanding the ecological consequences of desalination facilities and their water discharges on the marine ecosystem (Almasoudi & Jamoussi, 2024). The primary causes of these issues are chemical discharges and concentration, which can diminish seawater quality and impact marine life, energy requirements, air pollutants, and the consumption of processes, resulting in noise, visual disturbances, and land use (Bombar et al., 2016). It is necessary to present strategies for an environmental impact assessment (EIA) to ensure the establishment of environmentally sustainable projects, effective preventive and remedial actions, and environmental and social approval regarding desalination plants. This is because desalination plants have the potential to have adverse effects on the environment in coastal regions. Optimal planning can effectively reduce the majority of the environmental impacts (EIs) (Kress, 2019).

Saudi Arabia, South Africa, the United States, Israel, Dubai, and Australia are among the countries that have established primary desalination plants. Roughly 15,000 desalination facilities are distributed worldwide, with operational capacity varying from 50,000.0 m^3/day to 2,500,000.0 m^3/day. Table 9.1 presents a compilation of the ten largest desalination plants worldwide (Reuters Staff, 2020).

TABLE 9.1 World's largest desalination plants (Reuters Staff, 2020)

SR. NO.	*NAME OF THE PLANT*	*METHOD USED*	*CAPACITY (M^3/DAY)*	*LOCATION*	*REFERENCE*
1	Ras Al-Khair	Thermal multi-stage flash & reverse osmosis	1,036,000	Saudi Arabia	Esmaeilion (2020)
2	Taweelah	Thermal multi-stage flash & reverse osmosis	230,000	UAE	JICA (2016)
3	Jubail Water and Power Company	Multi-effect distillation	23,500	Saudi Arabia	JICA (2016)
4	Umm Al Quwain	Reverse osmosis	590,000	UAE	JICA (2016)
7	Sorek-1	Reverse osmosis	624,000	Israel	Nagaraja Bhat (2016)
8	Sorek-2	Reverse osmosis	570,000	Israel	Nagaraja Bhat (2016)
9	Fujairah-2	Reverse osmosis	136,000	UAE	JICA (2016)

Desalination is primarily categorized based on the technology and feedwater type, resulting in diverse effects across different methods. According to Heck et al. (2018) and Panagopoulos (2021b), brine discharges are the primary variable contributing to desalination's environmental implications. Brine releases can degrade coastal water quality and adversely harm marine life, posing a significant environmental threat (Elsaid, Sayed, et al., 2020a). The release of air pollutants is related to the energy requirements of the processes. Thus, it is necessary to assess the probable EIs of desalination facilities while minimizing any adverse consequences. Desalination's EIs have garnered significant interest as desalination is increasingly being applied in numerous environments to address critical water requirements (Liponi et al., 2020). In addition to the increased focus on environmental conservation initiatives, therefore mitigating the effects of human activities in overall and industrial operations (Caldera and Breyer, 2019). The primary goal of this focus is to mitigate or decrease these various environmental inputs from the origin by using highly efficient desalination methods and renewable energy sources (Muhammad Yaqub & Lee, 2019).

The environmental effects of desalination can be indirect or direct, depending on the circumstances (Dawoud, 2012). For instance, the following components of membrane desalination have the potential to impact the surroundings directly: the supply of feed water, the disposal of brine, the desalination process, the pretreatment procedure for feed water, and the energy requirements (Ihsanullah et al., 2021). The discharge of brine is predominantly categorized as impacting the environment due to its status as the liquid waste generated during the desalination process. The brine contains corrosive compounds, filter backwash, and pretreatment compounds (Lee & Jepson, 2021).

The indirect environmental pollution results from the necessity of enhancing energy generation for desalination facilities. Recent technological advancements, particularly in energy recovery systems, have improved energy efficiency in seawater RO desalination plants (Andersen & Bank, 2012). Desalination facilities require an external supply of electrical energy produced by thermal power plants. Electricity generation necessitates the combustion of fuels in a thermal power plant, releasing detrimental flue gases, including carbon dioxide, into the environment. The outcome of this mechanism is the phenomenon of global warming. Impacts on marine ecology resulting from restoring concentrated brine into the ocean (Petersen et al., 2018). Reverse osmosis (RO) desalination facilities often control the brine discharge flow rate from 30% to 70% of the feed water flow, yielding a concentration 1.2–1.6 times greater than seawater. It is critical to select an appropriate technology to distribute brine discharge effectively to lessen its detrimental influence on the marine environment (Meerganz von Medeazza, 2005). Climatic and hydro-geological factors inherent to the sea, including currents, bathymetry, waves, and water column depth, impact the extent to which the concentrated brine is released into the sea, which is the principal source of the effect. These parameters would define the extent to which the brines mixed and, by extension, the affected area's geographical extent (Elsaid, Kamil, et al., 2020a).

Ensuring compliance with environmental rules is crucial to building a desalination plant that can be approved (Younos & Lee, 2020). Seawater desalination differs from brackish water desalination in several respects, and these differences should be noted. When using seawater as a feed, the concentration of brine effluent is often lower than that of groundwater because the former is typically mixed with cooling water during the desalting (T Younos, 2005). According to Lattemann and Höpner (2008), most seawater desalination units are connected to power plants. As a result, this combined effluent is also mixed with a power plant's cooling water. Since this is the case, the temperature of this brine is correspondingly higher. Chemicals are added to control biofouling, foaming, scaling, and corrosion at seawater intake, which also goes into brine water. Sweater desalination effluents are the only ones that contain biocides and antifoaming chemicals. However, brackish feedwater is also treated with antiscalants because RO technology is typically employed for inland brackish water treatment rather than multi-stage flash (MSF) and multi-effect distillation (MED) technology for seawater; thermal desalination results in somewhat more significant greenhouse gas (GHG) emissions than brackish water desalination (Abdul-Wahab & Al-Weshahi, 2009).

This chapter aims to provide a comprehensive overview of desalination's health and EIs. It highlights the growing global population and the increasing demand for freshwater, which has led to the exploration of desalination as a viable solution. The chapter discusses various desalination technologies, their environmental consequences, and the importance of conducting EIAs to ensure sustainable development. In addition, it highlights the importance of appropriate planning and mitigation techniques to limit the adverse effects of desalination on aquatic environments and coastal conditions. This chapter aims to provide an equitable viewpoint on the advantages and disadvantages of desalination, with the goal of facilitating informed decision-making among prospective projects.

9.2 ENVIRONMENTAL AND HEALTH ISSUES

The desalination of Brackish or seawater has several advantages for economics, humankind, and the environment. It offers an inexhaustible source of quality drinking water without adversely affecting native freshwater ecosystems. However, there are concerns about the possible adverse effects (Dawoud, 2012). These are caused mainly by concentrate and chemical discharges, which may damage marine life and the cleanliness of the water along the coast, as well as the air pollution caused by the processes' need for energy. More effects might happen, but our information on sea discharges shows that all large projects must be thoroughly checked for EI (Hoepner & Lattemann, 2003). Desalination must be integrated into regional management strategies that regulate water supply usage and desalination technology to ensure that coastal regions develop sustainably and remain under control (UNEP/MAP/MEDPOL, 2003). In order to make desalination projects as eco-friendly as feasible, we need to ensure that the current effects are reduced as much as possible and weigh any worries about the future against the potential outcomes of other water sources and management approaches. This will help the technology last for a long time (Lattemann & Höpner, 2008) (Table 9.2).

Regularly discharging effluents from the power and desalination plant impacts the coastal environment. The species living in the discharge area are often impacted by the localized rise in seawater temperatures caused by water effluents. Lower quantities of

TABLE 9.2 Different parameter impacts (Elsaid, Kamil, et al., 2020b)

	PARAMETER IMPACTS	*TYPE OF IMPACTS*	*DESCRIPTION*
1	**Salinity**	Physical	Convective currents developed due to differences in density are responsible for alteration in the water parameter.
		Chemical	The salinity of the water increased.
			This causes an increase in the sediments' salinity due to the brine's higher salinity, which has a negative buoyancy.
			Reduces the solubility of oxygen, and as a result, the quantity of dissolved oxygen.
		Biological	Tolerance to saltwater causes the death of marine life.
			The benthic zone gets disturbed & leads to non-indigenous species.
			Metabolic, photosynthesis, and growth rates get changed.
2	**Disinfection byproducts**	Physical	Toxic compound degrades aquatic organisms.
		Chemical	Marine life dies from exposure to hazardous chemicals.
3	**Chemical additives (antiscalants, coagulants, flocculating agents, polyelectrolytes, etc.)**	Physical	Suspended solids increase, which ultimately increases turbidity, which further affects light penetration depth.
		Chemical	Harmful chemicals are added to the marine environment.
			Iron salts used as coagulants affect the light penetration depth of water.
		Biological	The benthic zone gets disturbed and leads to non-indigenous species.
4	**Heavy metals (Corrosion products etc.)**	Physical	Suspended solids increase, which ultimately increases turbidity, which further affects light penetration depth.
		Chemical	Corrugation increased.
		Biological	Heavy metal intolerance causes the death of marine life.
			Metabolic, photosynthesis, & growth rates get changed.

dissolved oxygen can result from heat-related changes to water quality processes. On top of that, chlorinating the cooling water could add harmful chemicals to the water. The salinity of the received water can be increased via desalination systems (Liu et al., 2013; Petersen et al., 2018).

Novel and more efficient desalination procedures have been developed due to technological developments. However, the management and processing of rejected brine, the principal byproduct waste of most desalination facilities, has not been improved (Gupta et al., 2003). Most plants face increasing costs and significant environmental issues when disposing or managing desalination brine (concentrate). Inland plants have had few options regarding brine management (Ahmed et al., 2002). This is even though this ecological and financial problem is of significant scope. These methods include evaporation ponds, deep well injection, mechanical/thermal evaporation, land disposal, and surface water or wastewater treatment plants for discharge. Potentially harmful to soil and groundwater, including anti-scale additives, reject brine comprises a wide range of chemicals and inorganic salts at varying amounts (Abdul-Wahab & Al-Weshahi, 2009).

Environmental pollution caused by human activities pollutes the coastal marine environment, which in turn causes changes in the physiochemical features of the environment as well as the marine life that live there. Pollutants can affect physiochemical conditions, reduce hydrodynamic conditions, cause organic enrichment, induce hypoxia, and, recently, result in brine discharge. These factors can all change the circumstances of the environment (de-la-Ossa-Carretero et al., 2016). The primary ecological problems related to brine disposal include increased salinity in receiving water bodies and soils and the detrimental effects of brine-containing elevated levels of dissolved solids on benthic marine organisms near the discharge area (Panagopoulos, 2021a).

Additionally, brine is typically released into the ocean as a byproduct of saltwater desalination, which is practically inevitable (Meerganz von Medeazza, 2005). The negative effects of brine on marine life can be caused by either a significantly elevated level of salt or by the introduction of unwanted contaminants that aren't present in the receiving body of water. The brine is quite salty but may contain other substances, such as acids, heavy metals, organic compounds, anti-fouling agents, chlorine, and dangerous preparation chemicals. These compounds are essential for treating the influent water and pipelines (Younos & Lee, 2020).

9.3 SEAWATER INTAKES AND PRETREATMENT

The inflow and outfall structures are accountable for removing brine and feedwater, respectively. The inflow structures are required for the withdrawal of feedwater. According to Jones et al. (2019), these facilities are essential because feed water is collected at a rate of around 250 million m^3/day, and brine is produced at a rate of 141.5 million m^3/day, which is 51.7 billion m^3/annum. There are a few different types of intakes: those above or below the water's surface, those located in wells or infiltration tunnels, and those located underground. "Open sea intake" encompasses both kinds of intake (Kress, 2019). Two distinct types of outfalls can be distinguished: open outfalls and submerged outfalls. The plant intake and outflow experience a crucial EI due to their direct interaction with the maritime environment (Kress et al., 2018). Entrainment, impingement, and entrapment (EI&E) are the main biological consequences associated with plant open sea intake. These impacts decrease the variety and quantity of aquatic flora in the marine environment close to the intake (Fariñas & López, 2007).

Seawater desalination plants might get their water in several ways, but the most typical is via open seawater intakes. The interaction of aquatic organisms with intake screens, obstructions, or their introduction into the facility via source water may lead to the decline of aquatic species due to the utilization of open intakes (Missimer & Maliva, 2018). The construction of the intake structure and pipelines initially disrupts the substrate, leading to the re-suspension of sediments, minerals, or pollutants into the water column. After they have been erected, the structures have the potential to influence sediment transport and water exchange, serve as artificial reefs for certain species, or interrupt shipping routes or other marine activities on the water.

Marine life faces a significant risk if there is a desalination surface water intake nearby because this type of intake is utilized to extract water (Petersen et al., 2018). Fish that have reached maturity, the larvae of fishes, and other marine species can become stuck or sucked into open water surface intake pipes, where they suffer severe injuries or death. One of the reports compiled by the State Water Resources Control Board found that one of the power plants in the state of California was responsible for the annual deaths of 70.0 billion fish larvae and other forms of aquatic life. This was due to the plant's open water surface intake. The brine waste produced is the most typical byproduct of desalination plants. This waste has a high concentration of salts and other compounds, which poses a considerable risk to marine life and water quality. Because it has a high density and salt content, it tends to collect at the point where it is released, smothering organisms that live on the ocean floor and dramatically affecting coastal ecosystems (Lattemann & Höpner, 2008). The following components are essential for the direct EIs of the membrane desalination processes (Figure 9.1).

Feedwater pretreatment primarily aims to prepare the water for the desalination processes that will come after it. Performing pretreatment in any desalination process is essential to ensure high-quality feed generation. This, in turn, improves plant productivity and operability by reducing the time that operations are interrupted and the number of times they are interrupted (Henthorne & Boysen, 2015). Colloidal and suspended particles are eliminated during pretreatment to prevent fouling and minimize scaling, leading to higher energy requirements for sustaining plant output (Fritzmann et al., 2007). The pretreatment process often involves coagulation, filtering, flocculation, and introducing numerous chemicals, including biocide scavengers, antifoams, acids, biocides, antiscalants, and corrosion inhibitors. Effective planning and optimizing feedwater pretreatment are crucial in reducing the EIs related to pretreatment since these chemicals will be released into the environment together with the brine stream (Yang et al., 2011). During pretreatment, the primary EIs are caused by biocides (Elsaid, Sayed, et al., 2020b). These biocides generate disinfection byproducts (DBPs) when they react with organic matter in source water. These DBPs have ecotoxic effects on marine life. Furthermore, according to Bar-Zeev et al. (2021), due to the elimination of aluminum and iron that contain high-sized particles due to coagulation and flocculation processes, there are some changes in color and turbidity.

Figure 9.1 Flowchart illustrating the components of membrane desalination, featuring a circular sequence: feedwater supply, pretreatment, desalination process, brine discharge, and energy requirements. Each step is in a distinct colored box linked by arrows.

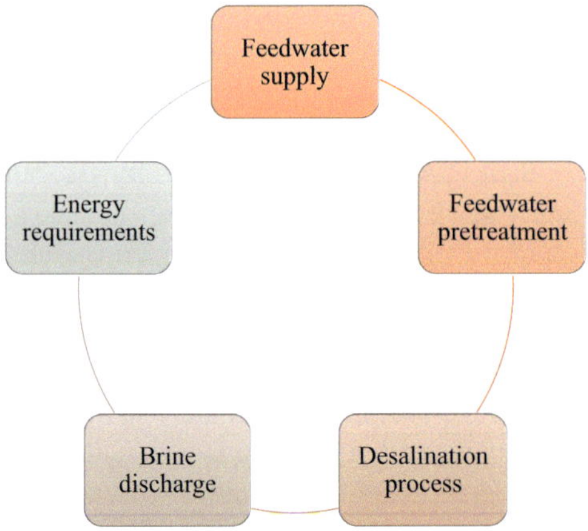

FIGURE 9.1 Components of membrane desalination.

9.4 REJECT STREAMS AND OUTFALLS (IMPACT OF BRINE DISCHARGE)

Concentrate brine is a byproduct of desalination operations. Sometimes brine can have a total dissolved solids (TDS) content higher than 36,000.0 mgL^{-1} (Mezher et al., 2011). In addition to anti-scale components, corrosion, and other reaction products, brine contains several compounds utilized for pretreatment and posttreatment (Dawoud, 2012). The recovery rate of the product water to feedwater ratio determines the total brine generated by a desalination plant. The plant's recovery ratio is directly related to the brine's salinity.

Desalination generates streams with elevated dissolved solids, including product water, freshwater, and concentrate water, often called brine. Therefore, brine is an undesired byproduct of the desalination unit and must be disposed of appropriately (Elsaid, Kamil, et al., 2020c). Within the context of brine discharge, the primary focus is on the quantity of brine water that is being discharged as well as the composition of the brine, which includes contaminants like heavy metals, both inorganic and organic components, and the byproducts of the pretreatment and posttreatment processes. Indeed, the disposal process is influenced by the composition of brine water (Mickley et al., 2006). The proportion of rejected brine in the feedwater ranges from 20% to 70%, depending on the feedwater's TDS concentration (Buros, 2000). The characteristics of the brine are directly influenced by the quality of the water supply, the desalination method employed, the chemical additives utilized, and the recovery ratio.

According to Mickley et al. (2006), the waste produced by the various components of membrane desalination can be classified into four distinct categories: pretreatment waste, cleaning waste, membrane concentration waste, and posttreatment trash. The pretreatment process for RO desalination may involve acidification, anti-scalant chemicals, chlorination, and dechlorination, among other possible procedures. Alternative

TABLE 9.3 Desalination plants' chemical pretreatment and posttreatment procedures (Sadhwani et al., 2005; T Younos, 2005)

PROCESS	*CHEMICAL USED*	*PURPOSE*
Pretreatment	NaOCl $FeCl_3/AlCl_3$ H_2SO_4/ HCl $NaHSO_3$ scale inhibitors	Removal of suspended solids from water, pH adjustment, inactivation of microorganisms, prevention of scale formation, neutralization of chlorine in the feedwater.
Posttreatment	Enzymes detergents, surfactants, and caustics biocides chelators acids	Dissolving of organic material and inactivation of microorganisms. Dissolving metals, silica, degradation of bacteria, scale removal, re-suspension of particulate matter.

pretreatment techniques, such as filtration, coagulation, ion exchange, and sorption, may be utilized in the case of feedwater that displays an insufficiently high level of quality. This is partly due to some offshore desalination facilities using groundwater as their feedwater possibly containing arsenic and radionuclides, which could contaminate the rejected brine (Sobhani et al., 2012). A list of the chemicals that are used in the pretreatment and posttreatment processes in desalination facilities is provided in Table 9.3.

Regarding both amount and composition, the brine produced as a byproduct of desalination procedures is considered the principal waste stream (Al-Othman et al., 2019). Regarding membrane desalination, the brine stream-to-product water stream ratio typically falls between 1 and 2 at 50–35% recovery rates (Henthorne & Boysen, 2015). This is consistent with the standard ranges observed in SWRO desalination. Several elements are primarily responsible for the effects of brine on aquatic ecosystems and marine life. Most EI studies concentrate on analyzing the features of the brine at the point where it is discharged and after it has been initially mixed with saltwater in the mixing zone. This allows for a more precise evaluation of the brine's impact (Mannan & Al-Ghamdi, 2019).

9.5 IMPACT OF DESALINATION ON THE ENVIRONMENT AND MITIGATION

9.5.1 Brine Management

The optimal, most convenient, and cost-effective method to release brine from coastline desalination facilities is releasing it into the ocean. The release of brine into the sea has a negative impact on the benthic ecosystem. Abdul-Wahab and Al-Weshahi (2009) state that the impact is conditional on the sea's waves, bathymetry, and water column depth, among other hydrological and environmental factors. Since desalination units are typically situated far from the ocean, it is not feasible to discharge these plants' brine into the ocean (Ahmed & Anwar, 2012). Desalination plants located inland confront the formidable challenge of brine disposal.

On the other hand, waste brine from inland desalination plants can be disposed of in various ways (Abdul-Wahab & Al-Weshahi, 2009). Some ways are well discharge, wastewater network discharge, surface disposal, injection into deep saline aquifers, and outfall disposal into the sea (El-Dessouky & Ettouney, 2002). Irrigating salt-tolerant plants and capturing brine shrimp are two more ways to manage brine (Ahmed et al., 2002). The problem is that these techniques don't deal with big amounts of salty water because of their fundamental limits (El-Dessouky & Ettouney, 2002).

According to Mohamed et al. (2005), it is vital to properly manage brine to avoid having negative effects on the environment that is receiving it. The disposal of brine is likely the most advantageous choice available in the modern period; However, the choice of disposal method is contingent upon several different criteria. According to Mezher et al. (2011), these factors include "the volume of concentrate, the properties of concentrate, the geographical site of the discharge point of concentrate, the availability of a receiving site, the permissibility of choice, approval from the public, capital, and operating costs, and the ability of the facility to be expanded." Mohamed et al. (2005) state that brine generated by desalination facilities located in coastal regions may be released into the ocean, provided that suitable outfall infrastructure is installed. Disposing brine by releasing it into a water stream is not a viable method because the change in salinity that might occur in inland water bodies can harm the ecosystem significantly (Williams et al., 2015). According to Mauguin et al. (2005), the release of brine directly onto land can lead to the salinization of the soil, which in turn destroys the soil. Evaporation ponds and the injection of brine into shallow wells may cause the contamination of groundwater resources. Meanwhile, from a shallow water table, the salt may rise to the water's surface through capillary action, eventually reaching the ground (Arnal et al., 2005).

Ayars et al. (2011) state that salt water can accelerate soil salinization, which reduces plant and soil production. This is because brine water has higher salinity and heightened chloride, sodium, and boron concentrations. Mohamed et al. (2005) suggest that concentrated brine can alter soil's electrical conductivity, directly affecting the Sodium Adsorption Ratio (SAR) and potentially inducing ion toxicity. A decrease in the permeability of soil occurs if there is an increase in SAR. In a nutshell, sodium causes the structure of the soil to change, making it more difficult for water to infiltrate the soil. This may affect the growth of plants. As a result of limited infiltration, runoff increases, oxygenation of the root zone is inadequate, and the soil's low permeability hinders salt leaching from the root zone. The latter causes the deposition of inorganic chemicals and heavy metals in the soil, which may have a more gradual impact (Alley, 2003). Additionally, brine infiltration may cause salt precipitation and alter the aquifer's hydraulic properties. The many EIs associated with desalination must be addressed through mitigation and control methods (M&CS) (Hoepner, 1999).

In order to decrease the amount of water, evaporation ponds are built to concentrate brine. However, the engineering design of evaporation ponds is related to environmental issues that must be considered (Khan et al., 2020; Ullah & Rasul, 2019). There is a

TABLE 9.4 Mitigation and control strategies (Elsaid, Sayed, et al., 2020c)

PARAMETERS	*EIS*	*MITIGATION STRATEGIES*
Pretreatment chemicals and heavy metals	Toxic DPB formation Increases water turbidity Metal salts used in pretreatment can cause color in the water. Contamination of groundwater aquifer.	Use the ZLD methods High recovery desalination Brine predilution with wastewater and cooling water Efficient pretreatment Use subsurface intake.
Higher temperature	Because of the rise in temperature at the point of discharge Reduces oxygen solubility and content Increase mortality of aquatic life	V Use of high-quality materials Proper maintenance plan Because of the rise in temperature at the point of discharge
Higher salinity	Increase salinity at the discharge point Salinity increases of sediments	Place the outfall in an active hydrodynamic zone with strong currents.

possibility that farmers whose farms are near these ponds will not be able to accept the utilization of this technology. There is also the possibility that water from the pond could seep into the fresh groundwater reserve through the structural breaks (Panagopoulos & Haralambous, 2020) (Table 9.4).

9.6 LIFE CYCLE ASSESSMENT OF DESALINATION PROCESSES

In line with the standards of ISO 14040 and ISO 14044, the utilization of life cycle assessment (LCA) methodology is one of the ways that may be utilized to evaluate the influence that a system has on the environment (Raluy et al., 2005). Raluy et al. (2004) suggest that LCA is a practical methodology for evaluating EIs across all stages of various human activities. Each life cycle evaluation considers a range of impacts, irrespective of their magnitude. The LCA analysis commences with the design and building of the desalination plant, followed by the diverse activities of water distribution and brine disposal associated with the facility (Mannan & Al-Ghamdi, 2019). LCA studies often include the following environmental effect categories: the potential for climate change, acidification, eutrophication, toxic effects on humans, and resources that are both renewable and nonrenewable (Zhou et al., 2011). These categories are included in the study because they are considered EIs. For the most part, previous assessments have only discussed LCA research on desalination from the perspective of environmental analysis and methodological approaches (Aziz & Hanafiah, 2021) (see Table 9.5).

TABLE 9.5 Summary of previous LCA study

SR. NO.	*PARAMETER STUDY FOR LCA OF DESALINATION*	*AUTHOR*
1	This study's primary focus was the LCA of several disposal options for utilizing RO membranes. He suggests conducting an in-depth assessment of the existing desalination markets and the environmental consequences of any potential future developments in the production and disposal of desalination membranes.	Lawler et al. (2015)
2	A study was conducted to evaluate the impact that several commercial desalination methods, such as MSF, SWRO, and MED, have had on the surrounding ecosystem. Additionally, he researched the incorporation of desalination into various other productive processes.	Raluy et al. (2005)
3	This study investigated LCA technique for SWRO process environmental performance metrics.	Vince et al. (2008)
4	This work conducted an in-depth study on the impact of desalination plants and brine disposal on the surrounding environment. As part of his research, they examined three potential solutions: direct disposal, dilution with the discharge from wastewater treatment plants, and dilution with salt water.	Meneses et al. (2010)
5	By analyzing their energy requirements and energy sources, compared the environmental effects of two desalination methods, sea water reverse osmosis (SWRO) and membrane-based distillation (MBD).	Aziz and Hanafiah (2021)
6	The environmental effects of multi-effect distillation, multi-stage flash, and RO were analyzed using LCA in connection with various forms of energy generation. Natural gas, coal, and oil are all potential alternatives to use as a source of energy.	Raluy et al. (2004)
7	Created a LCA-based impact assessment tool for analyzing the environmental effects of producing drinkable water.	Vince et al. (2008)
8	In order to examine the impacts of water importation, land reclamation, and sea desalination, the optimum sustainable water source was determined using a LCA method.	Lyons et al. (2009)
9	They analyzed the total air emissions produced by three distinct methods of water purification: desalination using traditional pretreatment, desalination using microfiltration (MF) and ultrafiltration (UF) membranes, and conventional pretreatment. Desalination of brackish groundwater and recycled wastewater.	Stokes and Horvath (2009)
10	Developed a LCA model of a combined water and sewage system. In-depth analyses also compare other possible futures, such as those in which the system's environmental performance is enhanced.	Lundie et al. (2004)

Assessing the effects on the environment, EIA is an important part of project planning and assessments, evaluating the many potential adverse environmental effects (Kariman et al., 2023). Since EIA may aid in determining the project site, execution, timetable, project plan, and many other essential factors, it is best to do it during the design process. The effects are rated according to the EIAR, or EIA report. If the linked EIs are within the authorities' tolerance levels, the project can move forward according to the established plans; otherwise, it can be evaluated with suggestions for mitigating and controlling the associated EIs; or it can be rejected altogether if EIs are severe (Rabaia et al., 2021).

Elsaid, Sayed, et al. (2020b) present the interactions that desalination technologies have with other crucial stages, such as brine disposal, pretreatment, energy usage, and intake/outfall, considering the environmental consequences (Figure 9.2).

Figure 9.2 Flowchart illustrating desalination technology and EIs, showing relationships between intake, pretreatment, recovery, and energy consumption. Arrows indicate effects on capacity, brine disposal, and EI, with color-coded influences and outcomes.

To determine desalination's effects on the surrounding environment, it is necessary to conduct EIA research. EIA research could determine the environmentally significant factors and assess the potential effects on the air, land, and groundwater (Ameen et al., 2018). The three main ways in which inland desalination plants affect the environment are as follows: first, by discharging concentrated brine, which can affect soil and

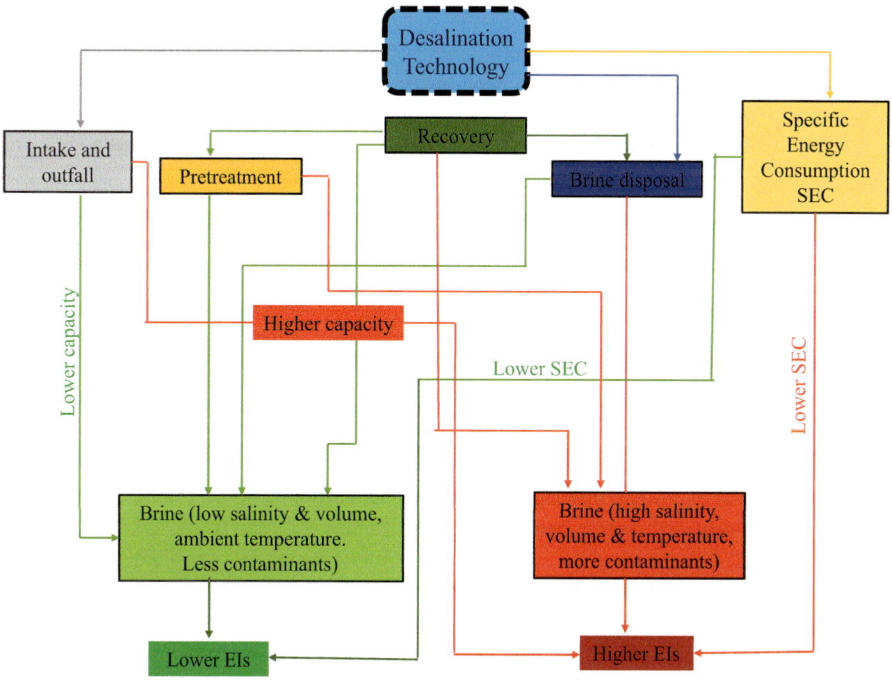

FIGURE 9.2 Desalination technology and EIs.

groundwater; second, by making noise with their powerful pumps and turbines; and, third, by polluting the air with greenhouse gases emitted from the fuel used to run the plant. An EIA study should aim to identify environmental damage and potential mitigation methods (Elsaid, Sayed, et al., 2020b).

The plant's location, the disposal of brine, and the amount of energy consumed are the three most significant environmental concerns related to a desalination plant (Son et al., 2020). The placement of the plant should be chosen with consideration given to the available energy supply, the distance to the feedwater intake, the position of the brine disposal, and the end-users. Certain circumstances, such as when a desalination plant is located in close proximity to residential areas, make it necessary to dispose of brine produced by land-based desalination plants (Campero et al., 2023).

The United Nations Environment Programme (UNEP) issued a report that provided direction about desalination in 2007 (UNEP 7). In their 2009 publication, Lattemann and Höpner (2008) offered a concise overview of the progression of the guidance document's development process and a summary of the most important findings. In order to carry out an EIA study of a desalination system, the UNEP publication suggests a technique that consists of ten steps. As illustrated in Figure 9.2, three principal phases are involved in these steps. The processes include project screening and scoping, EIA, and decision-making together with EIA review (Canton, 2021).

Mohammedi et al. (2013) conducted an EIA to examine the effects of the MSF process on aquatic life at Cape Djinet, Algeria. According to the findings of this study, the environmental consequences were mainly caused by the discharge of brine, with a lower degree of influence coming from the various chemicals utilized in the pretreatment process. The investigation considered four characteristics: temperature, pH, salinity, and conductivity. The findings indicated that there was no influence on the environment that was receiving the discharge and that all of the parameters that were measured in the discharge were in accordance with the Algerian standards that are in place.

Mustafa et al. (2018) carried out an EIA study to determine how heavy metals such as Cu, Fe, Cr, Ba, Zn, & Mn in the feedwater affected the efficiency of desalination facilities. For this study, the desalination facilities that will be evaluated include the Tobruk Desalination Plant and Benghazi North Power. The scaling and corrosion on all of the parts of desalination facilities due to precipitation were caused by metallic components in the feedwater stream, as the authors observed with their research. The presence of scale and corrosion in heat exchange pipes led to several operational issues, including a significant reduction in power consumption.

A study that was carried out by Shahabi et al. (2017) found that an Australian SWRO plant was examined utilizing two distinct scenarios for its feedwater. Two examples of such conditions were a beach well and an open sea inlet that used cartridge filtering before RO. Incorporating cartridge filtering into an open sea intake resulted in a 13% rise in carbon emissions, but the beach well intake scenario led to a 31% reduction in emissions. This was the conclusion presented by the researchers.

Environmental LCA studies on desalination technology have been conducted everywhere in the world. Most LCA studies were performed in Spain, accounting for 17 of the total 25 LCA studies (Aziz & Hanafiah, 2021). Because Spain is a water-stressed country and several of its regions are affected by water scarcity, it is Europe's largest consumer of Desalination (Aziz & Hanafiah, 2021).

Figure 9.3 Bar chart illustrating data from Environmental LCA studies on desalination technology across regions. Europe leads with a value of 25, followed by Asia at 17, the United States at 10, Australia at 9, North Africa at 5, South America at 3, and South Africa trailing with 2.

Mannan and Al-Ghamdi (2019) established a lifespan-based approach to assess the EI of MSF desalination on a broader scale. This framework was constructed about the lifespan. As stated by Mannan and Al-Ghamdi (2019), the methodology selected for this investigation is liable for determining the EI at each stage of the desalination procedure. Everything from the initial power consumption to the final disposal of chemical compounds and the desalination plant's functioning during its entire lifespan falls under this category. Researchers in Qatar measured the CO_2 emissions per cubic meter of desalinated water from three separate MSF desalination plants and found that they varied between 7.32 and 12.6 kg. This was revealed after the authors conducted their exploration. On the other hand, the LCA did not consider brine disposal when conducting the analysis (Almasoudi & Jamoussi, 2024).

According to Raluy et al. (2004), the LCA was utilized to investigate the impact of thermal (MSF, MED) and membrane (RO) desalination plants in Spain on the surrounding ecosystem. Based on the study's findings, the full life cycle of each desalination plant was considered, beginning with the raw materials, production/methods, operation,

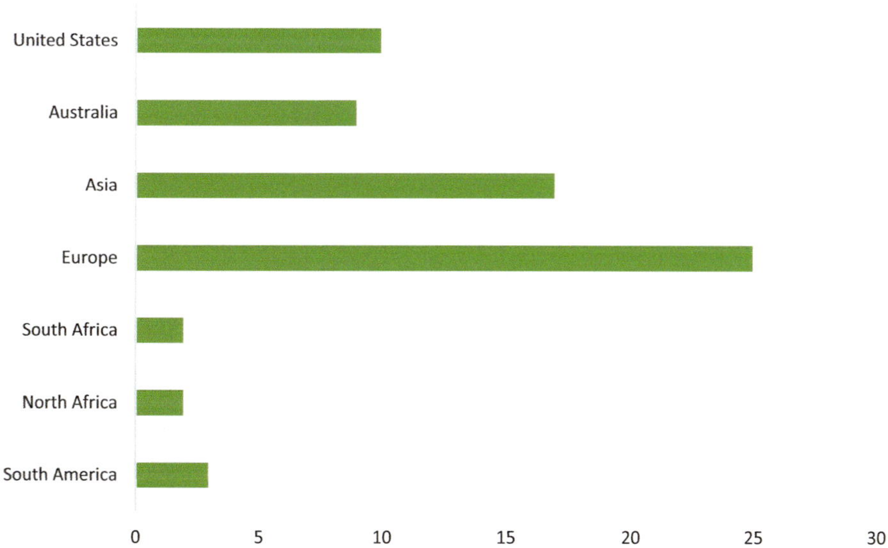

	South America	North Africa	South Africa	Europe	Asia	Australia	United States
■ Column1	3	2	2	25	17	9	10

FIGURE 9.3 Environmental LCA studies on desalination technology.

shipping, and disposal of reject brine. For the purpose of quantifying the environmental load, the software SimaPro 6.0 was utilized. MSF is responsible for emitting 23.39 kg of carbon dioxide, MED is responsible for emitting 18.04 kg of carbon dioxide, and RO is responsible for emitting 1.77 kg of carbon dioxide for every 1.0 m^3 of freshwater. The investigation also demonstrates that the operation phase is the most significant contributor since it is responsible for 96.98% of MED plants, 98.69% of MSF plants, and 99.05% of RO plants. Throughout the three stages of assembly, working, and final disposal, this is the step that makes the most notable contribution.

Shahabi et al. (2017) evaluated the use of renewable energy for a saltwater RO plant in the Australian city of Perth. The research generated an estimate of the emissions that happened during the desalination plant's construction and running phases using the software known as SimaPro. The three different scenarios for the generation of energy that were investigated were as follows: grid electricity that was solely produced from petroleum and coal, wind power, and a combination of 92% wind power and 8% photovoltaic solar power. When compared to the energy sourced from the grid, the data demonstrated that the desalination plant operated by sustainable energy had almost a 90% reduction in greenhouse gas emissions (Bello et al., 2021). In the scenario including grid electricity, electricity use during operation was responsible for 92% of the greenhouse gas emissions. However, renewable energy plants were the biggest contributors; chemical use during operation accounted for 60% of greenhouse gas emissions while building accounted for 17%. Between 36% and 39% of all greenhouse gas emissions came from indirect sources, such as chemical manufacture, photovoltaic solar panel manufacturing, and wind turbine manufacturing.

9.7 CONCLUSIONS

Desalination is an essential part of human progress because it solves the problem of water shortages in places that need it. Thermal and membrane filtration technologies are almost fully developed, meaning they can be used in real life and are technically and economically viable. New and developing desalination technologies are being worked on right now. These technologies will be more productive and use less energy than current desalination technologies. However, distillation has many environmental effects that need to be carefully examined. EIA is a big part of making decisions about desalination because it looks at how it will affect different parts of the environment.

Desalination with ZLD is the best way for freshwater desalination plants to deal with the brine they produce. Different ZLD strategies have been or are being developed, but more study and development are needed to solve technical, working, and economic problems. Inland Desalination faces two main environmental problems: how to handle the salt and how to reduce greenhouse gas emissions. The brine that is made when water is desalinated should be handled correctly. Various factors can alter the chemicals in brine: the quality of the feedwater, the distillation process employed, the recovery ratio, and the chemical additions utilized. Brine disposal methods in freshwater desalination plants include evaporation lakes, solar ponds, and deep well injection. Changing the soil's physical properties and potentially polluting groundwater are two environmental issues caused by applying brine water to land. Recently, efforts have been made to use brine more effectively and improve its management. However, more study and practical work needs to be done.

Another big problem is the GHGs that evaporation plants give off when they work. If you compare MSF and MED to RO, they make the most GHGs per m^3 of water. Several methods have been written about how to lower the amount of energy needed for purification. One method involves making an existing process better and more efficient. Something else is to switch to some new tools that might use less energy. A third way to do it is to use green energy sources to remove salt from the water.

REFERENCES

Abdul-Wahab, S. A., & Al-Weshahi, M. A. (2009). Brine management: Substituting chlorine with on-site produced sodium hypochlorite for environmentally improved desalination processes. *Water Resources Management*, *23*(12), 2437–2454. https://doi.org/10.1007/S11269-008-9389-7

Ahmed, M., & Anwar, R. (2012). An Assessment of the Environmental Impact of Brine Disposal in Marine Environment. *International Journal of Modern Engineering Research (IJMER)*, *2*(4), 2756–2761.

Ahmed, M., Shayya, W. H., Hoey, D., & Al-Handaly, J. (2002). Brine disposal from inland desalination plants research needs assessment. *Water International*, *27*(2), 194–201. https://doi.org/10.1080/02508060208686992

Alley, W. (2003). *Desalination of ground water: earth science perspectives*. https://books.google.com/books?hl=en&lr=&id=bIL6b_QMe0IC&oi=fnd&pg=PP3&dq=alley+2003+desalination&ots=NNsJdEA-d0&sig=CYqHUf-QIhHqW2rPWnQLfK914as

Almasoudi, S., & Jamoussi, B. (2024). Desalination technologies and their environmental impacts: A review. *Sustainable Chemistry One World*, *1*(November 2023), 100002. https://doi.org/10.1016/j.scowo.2024.100002

Al-Othman, A., Darwish, N. N., Qasim, M., Tawalbeh, M., Darwish, N. A., & Hilal, N. (2019). Nuclear desalination: A state-of-the-art review. *Desalination*, *457*, 39–61. https://doi.org/10.1016/j.desal.2019.01.002

Ameen, F., Stagner, J. A., & Ting, D. S. K. (2018). The carbon footprint and environmental impact assessment of desalination. *International Journal of Environmental Studies*, *75*(1), 45–58. https://doi.org/10.1080/00207233.2017.1389567

Andersen, I., & Bank, W. (2012). Environmental Impacts of Desalination. *Renewable Energy Desalination: An Emerging Solution to Close the Water Gap in the Middle East and North Africa*, 111–126.

Arnal, J., Sancho, M., Iborra, I., Gozálvez, J. et al. (2005). Concentration of brines from RO desalination plants by natural evaporation. *Desalination*. https://www.sciencedirect.com/science/article/pii/S0011916405004571

Ayars, J., Wallender, W. et al. (2011). On-farm irrigation and drainage practices. *Proceedings of the American Society of Civil Engineers*. https://www.ars.usda.gov/research/publication/?s/publication/??seqNo115=222995

Aziz, N. I. H. A., & Hanafiah, M. M. (2021). Application of life cycle assessment for desalination: Progress, challenges and future directions. *Environmental Pollution*, *268*, 115948. https://doi.org/10.1016/j.envpol.2020.115948

Bar-Zeev, E., Belkin, N., Speter, A., Reich, T., Geisler, E., & Rahav, E. (2021). Impacts of sewage outbursts on seawater reverse osmosis desalination. *Water Research*, *204*, 117631. https://doi.org/10.1016/j.watres.2021.117631

Bello, A. S., Zouari, N., Da'ana, D. A., Hahladakis, J. N., & Al-Ghouti, M. A. (2021). An overview of brine management: Emerging desalination technologies, life cycle assessment, and metal recovery methodologies. *Journal of Environmental Management*, *288*(August 2020). https://doi.org/10.1016/j.jenvman.2021.112358

Bombar, G., Dölgen, D., & Alpaslan, M. N. (2016). Environmental impacts and impact mitigation plans for desalination facilities. *Desalination and Water Treatment*, *57*(25), 11528–11539. https://doi.org/10.1080/19443994.2015.1089198

Buros, O. K. (2000). *The ABCs of desalting*. http://fac.ksu.edu.sa/sites/default/files/abcs_of_desalination.pdf

Campero, C., Bennett, N. J., & Arriagada, N. (2023). Technologies of dispossession in the blue economy: Socio-environmental impacts of seawater desalination in the Antofagasta Region of Chile. *Geographical Journal*, *189*(2), 231–245. https://doi.org/10.1111/geoj.12429

Canton, H. (2021). United Nations Environment Programme—UNEP. *The Europa Directory of International Organizations 2021*, 188–214. https://doi.org/10.4324/9781003179900-30

Dawoud, M. A. (2012). Environmental Impacts of Seawater Desalination: Arabian Gulf Case Study. *International Journal of Environment and Sustainability*, *1*(3). https://doi.org/10.24102/ijes.v1i3.96

de-la-Ossa-Carretero, J. et al. (2016). Response of amphipod assemblages to desalination brine discharge: impact and recovery. *Estuarine, Coastal and Shelf Science*. https://www.sciencedirect.com/science/article/pii/S027277141630035X

Younos, T. (2005). Desalination: Supplementing freshwater supplies approaches and challenges. *Academia.Edu*. https://www.academia.edu/download/100764172/60532467.pdf

El-Dessouky, H. T., & Ettouney, H. M. (2002). Fundamentals of Salt Water Desalination. In *Fundamentals of Salt Water Desalination*. Elsevier. https://doi.org/10.1016/b978-0-444-50810-2.x5000-3

Elsaid, K., Kamil, M., Sayed, E. T., Abdelkareem, M. A., Wilberforce, T., & Olabi, A. (2020a). Environmental impact of desalination technologies: A review. *Science of the Total Environment*, *748*. https://doi.org/10.1016/j.scitotenv.2020.141528

Elsaid, K., Sayed, E. T., Abdelkareem, M. A., Baroutaji, A., & Olabi, A. G. (2020b). Environmental impact of desalination processes: Mitigation and control strategies. *Science of the Total Environment*, *740*, 140125. https://doi.org/10.1016/j.scitotenv.2020.140125

Esmaeilion, F. (2020). Hybrid renewable energy systems for desalination. *Applied Water Science*, *10*(3), 1–47. https://doi.org/10.1007/s13201-020-1168-5

Fariñas, M., & López, L. Á. (2007). New and innovative sea water intake system for the desalination plant at San Pedro del Pinatar. *Desalination*, *203*(1–3), 199–217. https://doi.org/10.1016/j.desal.2006.01.036

Fritzmann, C., Löwenberg, J., Wintgens, T., & Melin, T. (2007). State-of-the-art of reverse osmosis desalination. *Desalination*, *216*(1–3), 1–76. https://doi.org/10.1016/j.desal.2006.12.009

Ghobeity, A., & Mitsos, A. (2014). Optimal design and operation of desalination systems: New challenges and recent advances. *Current Opinion in Chemical Engineering*, *6*, 61–68. https://doi.org/10.1016/j.coche.2014.09.008

Gupta, A., Gupta, S., & Patil, R. (2003). A comparison of water quality indices for coastal water. *Journal of Environmental Science and Health, Part A*. https://www.tandfonline.com/doi/abs/10.1081/ESE-120024458

Heck, N., Paytan, A., Potts, D. C., Haddad, B., & Lykkebo Petersen, K. (2018). Management preferences and attitudes regarding environmental impacts from seawater desalination: Insights from a small coastal community. *Ocean and Coastal Management, 163*, 22–29. https://doi.org/10.1016/j.ocecoaman.2018.05.024

Henthorne, L., & Boysen, B. (2015). State-of-the-art of reverse osmosis desalination pretreatment. *Desalination, 356*, 129–139. https://doi.org/10.1016/j.desal.2014.10.039

Hoepner, T. (1999). A procedure for environmental impact assessments (EIA) for seawater desalination plants. *Desalination, 124*(1–3), 1–12. https://doi.org/10.1016/S0011-9164(99)00083-1

Hoepner, T., & Lattemann, S. (2003). Chemical impacts from seawater desalination plants - A case study of the northern Red Sea. *Desalination, 152*(1–3), 133–140. https://doi.org/10.1016/S0011-9164(02)01056-1

Ihsanullah, I., Atieh, M. A., Sajid, M., & Nazal, M. K. (2021). Desalination and environment: A critical analysis of impacts, mitigation strategies, and greener desalination technologies. *Science of the Total Environment, 780*, 146585. https://doi.org/10.1016/j.scitotenv.2021.146585

JICA. (2016). *The Survey on Feasibility of Desalination Projects in Sub-Saharan Africa Final Report. June.* https://openjicareport.jica.go.jp/pdf/12264560.pdf

Jones, E., Smakhtin, V., Qadir, M., Van Vliet, M. T. H., Kang, S.-M., & Qadir, M. (2019). The state of desalination and brine production: A global outlook. *Science of the Total Environment, 657*, 1343–1356. https://doi.org/10.1016/j.scitotenv.2018.12.076

Kariman, H., Shafieian, A., & Khiadani, M. (2023). Small scale desalination technologies: A comprehensive review. *Desalination, 567*(September), 116985. https://doi.org/10.1016/j.desal.2023.116985

Khan, Q., Maraqa, M. A., & Mohamed, A. M. O. (2020). Inland desalination. In *Pollution Assessment for Sustainable Practices in Applied Sciences and Engineering.* https://doi.org/10.1016/B978-0-12-809582-9.00017-7

Koroneos, C., Dompros, A., & Roumbas, G. (2007). Renewable energy driven desalination systems modelling. *Journal of Cleaner Production, 15*(5), 449–464. https://doi.org/10.1016/j.jclepro.2005.07.017

Kress, N. (2019). Marine impacts of seawater desalination: Science, management, and policy. In *Marine Impacts of Seawater Desalination: Science, Management, and Policy.* https://doi.org/10.1016/C2016-0-01997-6

Kress, N., Oceanographic, I., & Galil, B. (2018). *Impact of seawater desalination by reverse osmosis on the marine environment Editor Query Form Book title: Author: Chapter: Efficient Desalination by Reverse Osmosis Nurit Kress and Bella Galil Query Please provide accessed date for all web address. R. April.*

Lattemann, S., & Höpner, T. (2008). Environmental impact and impact assessment of seawater desalination. *Desalination, 220*(1–3), 1–15. https://doi.org/10.1016/j.desal.2007.03.009

Lawler, W., Alvarez-Gaitan, J., Leslie, G., & Le-Clech, P. (2015). Comparative life cycle assessment of end-of-life options for reverse osmosis membranes. *Desalination, 357*, 45–54. https://doi.org/10.1016/J.DESAL.2014.10.013

Lee, K., & Jepson, W. (2021). Environmental impact of desalination: A systematic review of Life Cycle Assessment. *Desalination, 509*(October 2020), 115066. https://doi.org/10.1016/j.desal.2021.115066

Liponi, A., Wieland, C., & Baccioli, A. (2020). Multi-effect distillation plants for small-scale seawater desalination: thermodynamic and economic improvement. *Energy Conversion and Management, 205*(November 2019), 112337. https://doi.org/10.1016/j.enconman.2019.112337

Liu, T. K., Sheu, H. Y., & Tseng, C. N. (2013). Environmental impact assessment of seawater desalination plant under the framework of integrated coastal management. *Desalination, 326*, 10–18. https://doi.org/10.1016/j.desal.2013.07.003

Lundie, S., Peters, G. M., & Beavis, P. C. (2004). Life cycle assessment for sustainable metropolitan water systems planning. *Environmental Science and Technology, 38*(13), 3465–3473. https://doi.org/10.1021/ES034206M/ASSET/IMAGES/LARGE/ES034206MF00002.JPEG

Lyons, E., Zhang, P., Benn, T., Sharif, F., Li, K., Crittenden, J., Costanza, M., & Chen, Y. S. (2009). Life cycle assessment of three water supply systems: Importation, reclamation and desalination. *Water Science and Technology: Water Supply, 9*(4), 439–448. https://doi.org/10.2166/ws.2009.534

Mannan, M., & Al-Ghamdi, S. G. (2019). Life-cycle assessment of thermal desalination: Environmental perspective on a vital option for some countries. *World Environmental and Water Resources Congress 2019: Groundwater, Sustainability, Hydro-Climate/Climate Change, and Environmental Engineering - Selected Papers from the World Environmental and Water Resources Congress 2019*, 449–460. https://doi.org/10.1061/9780784482346.046

Mauguin, G., & Corsin, P. (2005). Concentrate and other waste disposals from SWRO plants: characterization and reduction of their environmental impact. *Desalination*. https://www.sciencedirect.com/science/article/pii/S0011916405004510

Meerganz von Medeazza, G. L. (2005). "Direct" and socially-induced environmental impacts of desalination. *Desalination, 185*(1–3), 57–70. https://doi.org/10.1016/j.desal.2005.03.071

Meneses, M., Pasqualino, J. C., & Castells, F. (2010). Environmental assessment of urban wastewater reuse: Treatment alternatives and applications. *Chemosphere, 81*(2), 266–272. https://doi.org/10.1016/J.CHEMOSPHERE.2010.05.053

Mezher, T., Fath, H., Abbas, Z., & Khaled, A. (2011). Techno-economic assessment and environmental impacts of desalination technologies. *Desalination, 266*(1–3), 263–273. https://doi.org/10.1016/j.desal.2010.08.035

Mickley, G. A., Hoxha, Z., Biada, J. M., Kenmuir, C. L., & Bacik, S. E. (2006). Acetaminophen self-administered in the drinking water increases the pain threshold of rats (Rattus norvegicus). *Journal of the American Association for Laboratory Animal Science, 45*(5), 48–54. https://www.ingentaconnect.com/content/aalas/jaalas/2006/00000045/00000005/art00007

Miller, S., Shemer, H., & Semiat, R. (2015). Energy and environmental issues in desalination. *Desalination, 366*, 2–8. https://doi.org/10.1016/j.desal.2014.11.034

Missimer, T. M., & Maliva, R. G. (2018). Environmental issues in seawater reverse osmosis desalination: Intakes and outfalls. *Desalination, 434*, 198–215. https://doi.org/10.1016/j.desal.2017.07.012

Mohamed, A., Maraqa, M., & Al Handhaly, J. (2005). Impact of land disposal of reject brine from desalination plants on soil and groundwater. *Desalination, 182*, 411–433. https://doi.org/10.1016/j.desal.2005.02.035

Mohammedi, K., Talamali, A., Smaili, Y., Saadoun, I., & Ait-Aider, A. (2013). Environmental impact of seawater desalination plants: case study in Algeria. *Researchgate.Net, 2*(6), 141–148. https://doi.org/10.11648/j.ajep.20130206.14

Muhammad, Yaqub, & Lee, W. (2019). Zero-liquid discharge (ZLD) technology for resource recovery from wastewater: A review. *Science of the Total Environment, 681*, 551–563. https://doi.org/10.1016/j.scitotenv.2019.05.062

Mustafa, M., Elabbar, S., Elmabrouk, F., Elabbar, M. M., & Elmabrouk, F. A. (2018). Environmental impact assessment for desalination plants in Libya. Case study: Benghazi North and Tobrouk desalination plants. *ElsevierMM Elabbar, FA ElmabroukDesalination, 2005•Elsevier*. https://doi.org/10.1016/j.desal.2005.02.074

Nagaraja Bhat. (2016). *Reverse Osmosis Monitoring & Optimization (ROMO): SAP HANA IoT Solution for Water Treatment | SAP Blogs*. https://blogs.sap.com/2016/03/15/romo-reverse-osmosis-monitoring-optimization-a-sap-hana-solution-for-reverse-osmosis-water-treatment/

Panagopoulos, A. (2021a). Energetic, economic and environmental assessment of zero liquid discharge (ZLD) brackish water and seawater desalination systems. *Energy Conversion and Management, 235.* https://doi.org/10.1016/j.enconman.2021.113957

Panagopoulos, A. (2021b). Water-energy nexus: desalination technologies and renewable energy sources. *Environmental Science and Pollution Research, 28*(17), 21009–21022. https://doi.org/10.1007/s11356-021-13332-8

Panagopoulos, A., & Haralambous, K. J. (2020). Environmental impacts of desalination and brine treatment - Challenges and mitigation measures. *Marine Pollution Bulletin, 161.* https://doi.org/10.1016/j.marpolbul.2020.111773

Petersen, K. L., Frank, H., Paytan, A., & Bar-Zeev, E. (2018). Impacts of Seawater Desalination on Coastal Environments. *Sustainable Desalination Handbook: Plant Selection, Design and Implementation*, 437–463. https://doi.org/10.1016/B978-0-12-809240-8.00011-3

Williams, P. M. et al. (2015). Technology for freeze concentration in the desalination industry. *Desalination.* https://www.sciencedirect.com/science/article/pii/S0011916414005463

Prajapati, M., Shah, M., Soni, B., Parikh, S., Sircar, A., Balchandani, S., Thakore, S., & Tala, M. (2021). Geothermal-solar integrated groundwater desalination system: Current status and future perspective. *Groundwater for Sustainable Development, 12.* https://doi.org/10.1016/j.gsd.2020.100506

Rabaia, M. K. H., Abdelkareem, M. A., Sayed, E. T., Elsaid, K., Chae, K. J., Wilberforce, T., & Olabi, A. G. (2021). Environmental impacts of solar energy systems: A review. *Science of the Total Environment, 754.* https://doi.org/10.1016/j.scitotenv.2020.141989

Raluy, R. G., Serra, L., & Uche, J. (2005). Life cycle assessment of desalination technologies integrated with renewable energies. *Desalination, 183*(1–3), 81–93. https://doi.org/10.1016/j.desal.2005.04.023

Raluy, R. G., Serra, L., Uche, J., & Valero, A. (2004). Life-cycle assessment of desalination technologies integrated with energy production systems. *Desalination, 167*(1–3), 445–458. https://doi.org/10.1016/j.desal.2004.06.160

Reuters Staff. (2020). *Inside the world's biggest water desalination plants | Reuters.* Reuters. https://www.reuters.com/article/us-saudi-water-desalination-idUSKBN26Y1HD

Sadhwani, J. J., Veza, J. M., & Santana, C. (2005). Case studies on environmental impact of seawater desalination. *Desalination, 185*(1–3), 1–8. https://doi.org/10.1016/j.desal.2005.02.072

Shahabi, M., McHugh, A., Anda, M., & Ho, G. (2017). A framework for planning sustainable seawater desalination water supply. *Science of the Total Environment.* https://www.sciencedirect.com/science/article/pii/S0048969716320678

Sirota, R., Winters, G., Levy, O., Marques, J., Paytan, A., Silverman, J., Sisma-Ventura, G., Rahav, E., Antler, G., & Bar-Zeev, E. (2024). Impacts of Desalination Brine Discharge on Benthic Ecosystems. *Environmental Science and Technology, 58*(13), 5631–5645. https://doi.org/10.1021/acs.est.3c07748

Sobhani, R., Abahusayn, M., Gabelich, C., & Rosso, D. (2012). Energy footprint analysis of brackish groundwater desalination with zero liquid discharge in inland areas of the Arabian Peninsula. *Desalination.* https://www.sciencedirect.com/science/article/pii/S001191641200080X

Son, H. S., Shahzad, M. W., Ghaffour, N., & Ng, K. C. (2020). Pilot studies on synergetic impacts of energy utilization in hybrid desalination system: Multi-effect distillation and adsorption cycle (MED-AD). *Desalination, 477*(October 2019), 114266. https://doi.org/10.1016/j.desal.2019.114266

Stokes, J. R., & Horvath, A. (2009). Energy and air emission effects of water supply. *Environmental Science and Technology, 43*(8), 2680–2687. https://doi.org/10.1021/ES801802H/SUPPL_FILE/ES801802H_SI_001.PDF

Ullah, I., & Rasul, M. G. (2019). Recent developments in solar thermal desalination technologies: A review. *Energies*, *12*(1). https://doi.org/10.3390/en12010119

Vince, F., Aoustin, E., Bréant, P., & Marechal, F. (2008). LCA tool for the environmental evaluation of potable water production. *Desalination*, *220*(1–3), 37–56. https://doi.org/10.1016/J.DESAL.2007.01.021

Voutchkov, N., Kaiser, G., Stover, R., Lienhart, J., & Awerbuch, L. (2019). Sustainable Management of Desalination Plant Concentrate-Desalination Industry Position Paper–Energy And Environment Committee of the International Desalination Association (IDA). In *The International Desalination Association World Congress on Desalination and Water Reuse* (pp. 1–32). UAE REF: IDAWC19-Voutchkov.

Yang, J., Sohn, J., & Kang, D. (2011). Market evaluation of seawater desalination plant considering international water scarcity and expense outlook by use and nation. *Journal of Korean Society on Water Environment*, *27*(2), 178–187. https://koreascience.kr/article/JAKO201107741472306.page

Younos, T., & Lee, J. (2020). Desalination: Concept and system components. *Corrosion and Fouling Control in Desalination Industry*, *3–27*. https://doi.org/10.1007/978-3-030-34284-5_1

Zhou, J., Chang, V. W. C., & Fane, A. G. (2011). Environmental life cycle assessment of reverse osmosis desalination: The influence of different life cycle impact assessment methods on the characterization results. *Desalination*, *283*, 227–236. https://doi.org/10.1016/j.desal.2011.04.066

Marine Environmental Impacts of Desalination Plants

10

10.1 INTRODUCTION

As a result of the growing need for freshwater on a global scale, desalination technology has emerged as a viable alternative to conventional methods of collecting freshwater (Goh et al., 2021). This technology has the potential to assist in addressing the issue of water scarcity on a global scale. According to Sun et al. (2021), it is projected that climate change will have an impact on hydroclimatic conditions, which will ultimately lead to a scarcity of water. The world's population growth and the economy's expansion are the key contributors to the worldwide water shortage. However, climate change is also anticipated to change the hydroclimatic conditions. On the other hand, most freshwater production occurs in the Arabian Gulf, along the coast of the Mediterranean, and near the Red Sea (Jones et al., 2019). This is even though the number of desalination projects has significantly expanded in all central locations worldwide over the past decade (Chandrasekharam et al., 2016).

Desalination, one of the many facets of the blue economy, has received very little attention from social scientists. According to Swyngedouw and Williams (2016), desalination has been seen as a sustainable method for enhancing freshwater sources while simultaneously resolving socio-environmental concerns resulting from the overuse of natural freshwater. Desalination systems have been increasingly prevalent worldwide due to the ever-present and frequently growing need for freshwater and the reduction in the economic cost brought about by several technological advancements. Gao et al. (2021) and Williams' (2022) study showed that around 15,906 desalination plants produced Brine at 141.5 million m^3 daily. These plants were responsible for pumping out more than 95.37 million m^3 of water daily. There is a need for increasing attention to be paid to the governance processes and socio-environmental downsides of desalination

DOI: 10.1201/9781003530893-10

(Swyngedouw & Williams, 2016). Desalination involves incorporating new stakeholders into ocean governance, reorganizing ownership structures regarding freshwater resources and the marine ecosystem, and generating distinct socio-environmental effects for local inhabitants (Aviram et al., 2014).

Wilder et al. (2016) examine desalination, which has concentrated on the environmental consequences, such as brine discharge and CO_2 emissions, as well as the impacts on livelihoods, water governance, and inequality. The results of this research have thrown light on how desalination transforms water into a commodity with environmental and social repercussions that can be easily justified when it is used to promote other forms of economic development, such as tourism, agriculture, and mining (McEvoy & Wilder, 2012). The unintended social effects include unequal distribution of the freshwater created (e.g., to industry instead of drinking) and higher water prices (Meerganz von Medeazza, 2005). On the other hand, comparatively limited research has been conducted on the socio-environmental effects of desalination (Fragkou, 2018).

Although desalination provides numerous benefits for human access to potable water, it is evident that, such as any other industry or technology, it may include various environmental and socio-economic repercussions that warrant careful consideration. According to Panagopoulos (2021), one of the most severe environmental effects of desalination is the discharge of the rejected brine stream. This hypersaline stream may contain harmful compounds like H_2SO_4, $AlCl_3$, NaOCl, and $FeCl_3$ from various desalination systems. Brine is frequently released into the marine ecosystem, and as a consequence, a variety of environmental concerns have surfaced as a consequence of this widespread practice. Furthermore, the desalination business is highly energy-intensive, and the application of fossil fuels is the primary energy source for this industry. According to Aleisa and Al-Shayji (2018), this energy source's adverse effects include releasing greenhouse gases (GHGs) and polluting the air. The operation and construction of desalination plants can have an impact on the surrounding environment, which is something that should be brought to our attention. There is also a significant degree of dependence on the plant's location and the source of the feed water sources (Al-Shayji & Aleisa, 2018).

Most of the impact on the marine environment occurs in the region surrounding the pipelines discharging concentrated Brine. The high specific weight of the concentrated Brine leads it to sink to the sea floor without being mixed first, even though it contains natural marine elements (Safrai & Zask, 2008). In addition, the marine life close to the pipe's exit may be harmed by the chemicals added to the water during the pretreatment stages of the desalination process. In addition, the physical positioning of the discharge pipelines may cause harm to marine communities that are particularly vulnerable (Ahmed & Anwar, 2012).

This chapter investigates the environmental impacts that desalination systems have on the ecosystem of maritime environments. It sheds light on the expanding significance of desalination technology in tackling the global water shortage issue caused by population growth, economic development, and climate change. This chapter aims to provide a more in-depth understanding of the intricate interactions between desalination procedures and the marine environment.

10.2 OVERVIEW OF ENVIRONMENTAL FOOTPRINTS OF DESALINATION

An ever-increasing number of desalination facilities are being constructed worldwide, which has led to an increase in the number of people concerned about the negative impact of these facilities on the environment (Loganathan et al., 2019). According to Ihsanullah et al. (2021), the most significant negative consequences of desalination on the environment are those associated with brine contamination, greenhouse gas emissions, intensive energy use, noise pollution, land utilization, and construction-related problems. For example, by reducing the capacity of the intake and outfall as well as the volume of Brine, it will be possible to obtain the best possible recovery, which will reduce the appropriate environmental consequences (Esrafilian & Ahmadi, 2019). Similarly, a desalination system with a high-energy performance can be indicated by a lower specific energy usage, reducing the environmental effects of energy consumption, such as greenhouse gas emissions. It is obvious that depending on the kind of desalination technology, pretreatment requirements would be severely altered as, in comparison with membrane-based desalination procedures, thermal-based desalination requires less pretreatment and, in turn, significant ecological effects (Elsaid, Sayed, et al., 2020a).

10.2.1 Effects on the Marine Ecology

It is generally agreed that the most significant undesired disposal produced by the various desalination technologies is the substantial amount of reject stream. This rejects stream discharge from thermal desalination systems, which often have a high temperature and contain a variety of chemical wastes that are applied throughout the desalination process. These wastes include solvents used for antifoaming, anticorrosion, and anti-scaling purposes, in addition to various side-products and heavy metals. According to Mauguin and Corsin (2005), these disposals can potentially adversely affect the organisms inside the marine community. Therefore, they have the potential to produce adverse changes in the physiological conditions (Frank et al., 2019), metabolic rates, and diversity of living organisms, as well as the physicochemical features of the environment, such as the level of dissolved oxygen (DO), turbidity, temperature, salinity, and chemical composition (Kress, 2019). The availability of heavy metals in the disposal sites is mainly associated with thermal desalination systems. These systems, which are subject to substantial temperatures during the process, have the potential to produce corrosion in a variety of metallic facilities. In particular, heavy metals such as cobalt and nickel may be present in pumps and heat exchangers when their alloys are subjected to wear and tear (Panagopoulos, 2022; Panagopoulos & Haralambous, 2020). However, the negative impacts of these heavy metals in membrane-based systems such as FO (forward osmosis) and RO are often below critical levels. This is because non-metallic materials like polymers are primarily utilized to design these systems (Figler et al., 2019).

Negative impacts can happen to the marine environment, especially when sensitive ecosystems are near areas with a lot of wastewater discharge. When it comes to the marine environment, the effects of a desalination plant depend not only on the physicochemical characteristics of the reject streams but also on the hydrographical and biological characteristics of the environment receiving the produced water (Roberts et al., 2010; Sola et al., 2020a). When compared to exposed, high-energy, open-sea environments, which are more capable of diluting and dispersing the discharges from desalination plants, it is commonly considered that enclosed and shallow sites with numerous marine life are more sensitive to the discharges from desalination plants (Hoepner, 1999). There is a substantial relationship between the physicochemical parameters of the discharges, the desalination process, and the pretreatment used.

Most large-scale sea water reverse osmosis (SWRO) desalination plants release their Brine into the coastal environment through surface discharge, which involves opening beach channels or offshore diffuser systems that use underwater nozzles (Rybar et al., 2005). Several elements, including discharge volumes, coastal bathymetry, dilution factors, discharge technology, brine and seawater densities, water currents, and tidal and wave activity, all have a role in determining the region affected by the brine that is discharged. Most research evaluating desalination brine's effects on the marine environment has concentrated on the visible mixing zone, where the salts and additives are substantially concentrated to be readily measured or induce a physiological reaction in marine organisms (Mauguin & Corsin, 2005). However, recent modeling initiatives have indicated that SWRO brine could have a far-field effect, affecting extensive coastal water regions. After the brine release into the coastal zone, hydrodynamic model simulations have shown that although not physiologically significant for the organisms involved, the diluted plumes may impact the far-field dynamics. Using simulated brine flow, it was demonstrated that gravity currents and the Coriolis force would cause the dense plume to expand over the seabed toward the slope (Missimer et al., 2018).

As desalination facilities grow more critical, it is essential to analyze and synthesize data on brine discharge's environmental and ecological effects on marine environments (Roberts et al., 2010). Many literature reviews have quantified Brine's physical, chemical, and biological effects on marine environments. Brine discharge has several possible implications, but few studies have examined its influence on marine communities around discharge outlets (Almasoudi & Jamoussi, 2024). Many desalination plants have been built worldwide, but little is known about their discharge effects (Bombar et al., 2016; Dawoud, 2012). This review examines Brine's physiochemical and ecological effects on seawater temperature, alkalinity, salinity, DO, nutrients, and chemicals, which harm the marine environment and highlight knowledge gaps.

10.2.1.1 Impingement and Entrainment

Some desalination plants are required to take in a significant amount of seawater, which can harm marine life in two different ways: impingement and entrainment through the utilization of seawater. Impingement is the collision with the screen, and entrainment is the drawing into the plant (Lattemann & Höpner, 2008c). Impingement indicates that the plant is being drawn in. Impingement is the process by which creatures are drawn into an intake pipe and then trapped in a fish screen that covers the intake, resulting in harm or death of the organisms. Entrainment takes place when small creatures pass through the fish screen and are subsequently transported into the intake pipes (Andersen & Bank, 2012). On the other hand, smaller organisms such as phytoplankton and zooplankton can be impinged or entrained by them, as demonstrated by several studies that were conducted in the past. They are not dangerous for large species because of the low flow rate.

10.2.1.2 Temperature

Brine discharge can have many significant effects, one of the most significant of which is thermal pollution, which can occur when the temperature of seawater is raised (Safrai & Zask, 2008). The distribution and natural equilibrium of marine flora and fauna species have been the subject of many studies (Danoun & Jones, 2007). These studies have been conducted to discover how these species react to shifting temperatures. According to the findings of these investigations, there is a direct connection between temperature and marine species. It is not obvious whether the ranges of temperatures that were examined because of desalination were based on field measurements or model projections (Kress, 2019). The temperature of the Brine that has been released is significantly higher than the temperature of the seawater in the environment. According to Danoun and Jones (2007), the temperature of seawater typically ranges from 10°C to 25°C. However, it will increase by around 60°C to 40°C near the saline disposal area. The method of temperature change is primarily determined by the type of treatment plants that are being used. Compared to reverse osmosis (RO), distillation units are responsible for a more significant increase in temperature. Abdul-Wahab and Abdo (2007) found a direct correlation between the temperature of the saltwater and the distance from the discharge location. The location of the plant discharge is another factor that determines the distribution and extent of the temperature change. According to Roberts et al. (2010), brine discharges that occur in a well-cleansed environment are more cost-effective and efficient than brine discharges that occur close to bodies of water.

Because membrane desalination requires a minimal temperature increase, it is essential to determine that the effects of Brine's temperature and pH are far less significant (Keirstad, 1962). Moreover, it has been noted that the brine and feedwater temperatures are pretty near, with a maximum difference of 2°C. This could be due to heat loss from high-pressure pumps or friction in the RO module passages (Sola et al., 2020a).

10.2.1.3 Density

The reject streams of RO plants maintain a greater density than saltwater because of their elevated salt concentrations (Danoun & Jones, 2007). Without absorption via a diffuser system, the discharge plume will fall to the sea floor in shallow waters and spread over the seabed. Regarding multi-stage flash (MSF) plants, it is necessary to consider the impact of temperature on density. Given that both parameters have opposing impacts on density, the distillation discharge can exhibit positive, neutral, or negative buoyancy depending on the atmospheric density stratification. However, in most cases, it will be neutral or positive buoyant owing to the more significant influence of temperature (Al-Dousari et al., 2012).

10.2.1.4 Dissolve Oxygen

The amount of DO in the ocean is also changed by dumping Brine. Hypoxia happens since the amount of DO decreases as salt increases (Haurwitz et al., 2008). Hypoxia is dangerous when there is not enough DO in the water. It can do much damage to aquatic life. A study from the California Coastal Commission in 2004 says that if the water temperature rises because of Brine, it can lower the amount of DO and raise the biomass of organisms suspended in the water (California Coastal Commission, 2004). In RO, the temperature does not change much, so warmth probably will not result in significant DO variations. Mixing chemicals that use oxygen in RO plants to stop rusting and remove leftover chlorine can lower the DO (Lattemann & Höpner, 2008a).

10.2.1.5 Salinity

The influence of Brine on seawater salinity and its consequences for the marine ecosystem has been detailed in several sources. According to Roberts et al. (2010), the concentration and expansion of brine plumes on the receiving water is the primary focus of analysis in such research. Since brine discharge produces a hypersaline layer that lowers toward the seabed due to its greater density, it is now commonly acknowledged that it has the potential to significantly impact marine life in the surrounding area (Meerganz von Medeazza, 2005). Alterations in salinity affect how marine species grow, affecting the rate at which they develop and increase. Larval stages are crucial transition times for marine species, and increased salinity causes considerable disruption to that period (Neuparth et al., 2002). Although marine life that lives in salty water is used to the fluctuation of salinity concentrations, it is possible that they will not be able to survive the sudden rise in salinity due to brine disposal. Because released Brine is significantly denser than saltwater, brine plumes tend to spread deeper along the bottom than they do along the surface (Haurwitz et al., 2008). This factor leads to the increased exposure of benthic animals to brine discharges compared to pelagic and planktonic organisms (Roberts et al., 2010). Even though RO technologies do not significantly impact temperature change, brine discharges from these processes have been shown to raise the salinity by a factor of two compared to the salinity of seawater (Tularam & Ilahee, 2007). According to the findings of several studies, the distance between the plant site and the brine discharge indicates a clear correlation between the excess salinity level of seawater and the brine discharge. It demonstrates that increasing the salinity level to a level of 50 parts per thousand might most likely have a significant impact on the size of the fish as well as their rate of survival (Heck et al., 2018).

10.2.1.6 Other Aspects

Brine discharge has several significant effects, one of which is high total alkalinity, which causes the amount of calcium carbonate, calcium sulfate, and other components in the seawater to increase by a factor of two compared to its usual level. Very few tests have been conducted to assess the change in total alkalinity in seawater caused by brine disposal, as well as the limit of tolerance that marine life has for Brine (Danoun & Jones, 2007). According to Sirota et al. (2024), brine discharge also causes changes in the pH range of marine ecosystems; however, these changes are highly insignificant compared to the other impacts.

The pollution of coastal ecosystems by the excessive release of brine wastewater and other chemicals that are utilised in desalination. Among these substances are various antiscalants and coagulants. By adding organic phosphorus (e.g., polyphosphates), antiscalants can cause oligotrophic coastal areas to become locally eutrophic, while coagulants can make the water more turbid and discolored (Sola et al., 2020b). Desalination plants rely on pretreatment and posttreatment processes that add chemicals to keep membranes from deteriorating or improving flocculation (Frank et al., 2019). Mermaids and other marine life in the area near the outlet are affected by the chemical components employed, such as anti-scaling agents (polyphosphates, polymers of maleic acid, sulfuric acid, etc.) and antifoaming agents (fatty acids, alkylated polyglycols, etc.), which are discharged with Brine (Meerganz von Medeazza, 2005). The environmental effects of brine disposal have been extensively studied using laboratory experiments and field monitoring.

10.3 PRETREATMENT AND CHEMICAL PROPERTIES OF REJECT STREAMS

To improve plant efficiency and increase intervals between shutdowns for cleaning or component replacement, chemical pretreatment of the intake water is necessary in most desalination plants (Elsaid, Kamil, et al., 2020). Biological fouling, scaling, and corrosion are especially common in thermal plant heat exchanger systems that come into contact with brine and salt water. The membrane is the weakest link in an RO plant because it allows water to pass through but not most dissolved and suspended compounds (Sadhwani et al., 2005). These compounds can pool on the membrane, leading to "fouling" from microorganisms, suspended particles, or scale deposits. Controlling biofouling and scale is a standard pretreatment step in both plants (Mezher et al., 2011). Distinguishing features include RO plants' ability to remove suspended materials and distillation plants' ability to manage corrosion and foaming. The residues and waste products of cleaning and pretreatment chemicals are frequently released into the water along with the concentration. The plant's intake, saltwater quality, the desalination process, and the pretreatment and cleaning scheme are the primary determinants of the concentrate's composition (J. Zhou et al., 2011).

At the plant's intake, chlorine is frequently used as either chlorine gas or hypochlorite. Desalination plants often use chlorine doses that vary between 0.5 mg/L and 2 mg/L, comparable to the treatment frequently used for cooling water supply in power plants (Elsaid, Sayed, et al., 2020b). As a result of self-decomposition and the oxidant demand of the seawater that is being taken in, the desalination process results in the loss of the majority of the oxidative capacity. Before the feedwater enters the RO units, residual oxidative capacity is often neutralized, frequently with sodium bisulfite, to prevent membrane damage. This is done since most RO plants work on polyamide membranes sensitive to oxidizing agents (Ameen et al., 2018).

As a consequence, the majority of RO discharges are characterized by levels of residual oxidants that are extremely low or non-detectable. On the other hand, non-oxidizing biocides can be utilized and released into the environment alongside the brine (Elsaid, Sayed, et al., 2020a). One of the most significant concerns regarding distillation plant reject streams is the presence of residual chlorine levels. This is because chlorine removal is typically not required in such facilities unless environmental requirements mandate it. The intake water is used to cool in thermal plants, accounting for approximately two-thirds of the total, while only one-third is used as feedwater for desalination (Hoepner, 1999).

It has been reported by Lattemann and Höpner (2008a) that the discharge levels of residual oxidants range between 0.2 mg/L and 0.5 mg/L. It is necessary to provide occasional shock treatment when chlorination is used, which is a drawback. The decomposition of organic matter by chlorine results in the formation of smaller fragments, which in turn offer nutrients to microorganisms that are able to survive and frequently lead to heavy regrowth. Chlorine concentrations are raised periodically, as frequently as once or many times per day, with chlorine doses varying between 6 mg/L and 8 mg/L in distillation units (Ghaffour et al., 2013; Lattemann & Höpner, 2008d). This is done in order to control regrowth. As a result of the environmental issues brought about by residual chlorine and the by-products of disinfection, several different neutralization and alternative pretreatment procedures have been considered. There are a variety of chemicals that can be utilized for neutralization; however, sodium bisulfite is the one that is utilized in RO plants the most frequently. Some alternatives, such as hydrogen peroxide and sulfur dioxide, have been proposed as potential treatments for the reject streams produced by thermal plants (Chen et al., 2019; Missimer et al., 2018).

10.3.1 Coagulants and Coagulant Aids (RO Plants)

Coagulants such as ferric- or aluminum chloride are used to enhance the filtering of suspended material from the RO feedwater (Petersen, Paytan, et al., 2018). To improve coagulation, coagulant aids, which are organic compounds with large molecular weights such as polyacrylamide, can be employed. The pH of the feedwater is usually regulated using sulfuric acid. A relatively low harmful potential characterizes coagulants and coagulant aids (Omerspahic et al., 2022). However, the release of solid material might change the reject stream's color when ferric salts are used, causing a rise in turbidity that may decrease light penetration and, hence, primary production or lead to the burial of stationary benthic organisms (Roberts et al., 2010).

Ferric chloride, aluminum sulfate, and polyelectrolyte are commonly used as coagulants to eliminate suspended solids from saltwater (Kennedy & Jacoby, 1999). Traditionally, these substances were eliminated using multimedia filters followed by cartridge filters (Fritzmann et al., 2007; Shenvi et al., 2015). In recent years, ultrafiltration (UF) and nanofiltration (NF) have been extensively used as pretreatment methods for RO desalination. They aim to enhance feedwater quality and minimize RO membrane fouling (Ma & Lu, 2011; Voutchkov, 2020; Voutchkov et al., 2019). SWRO desalination typically involves the addition of ferric chloride and polyelectrolyte at dosages of 1 to 5.2 and 0.2–1.4 ppm/m^3, respectively (Khawaji et al., 2007, 2008). An antiscalant is often included to manage scaling by adding sparingly soluble salts, such as magnesium and calcium carbonates and barium, calcium, and strontium sulfates. These measures assist in preserving plant productivity and minimizing periods of inactivity, particularly during periods of enhanced recovery (Fritzmann et al., 2007; Sharaf et al., 2011). The recommended dosage of antiscalants is typically between 1 ppm/m^3 and 3 ppm/m^3 and is determined by an antiscalant formulator based on the quality of the feedwater and the intended recovery rates (Fritzmann et al., 2007).

10.3.2 Heavy Metals

Due to internal corrosion within the plants, metals might be introduced into the reject streams. The use of copper-nickel alloys is prevalent in heat exchanger surfaces inside distillation plants, while the issue of copper contamination in Brine is often faced in thermal desalination operations (Kress, 2019). RO desalination plants predominantly use non-metal equipment and stainless steel. The RO brine may include small amounts of iron, chromium, nickel, and molybdenum. However, the contamination with these metals is typically not below a critical threshold in facilities primarily using stainless steel (Al-Dousari et al., 2012).

Heavy metal contamination in RO desalination plants is often below a critical threshold since the primary materials used are non-metal polymeric components and stainless steels (Tularam & Ilahee, 2007). D. Zhou et al. (2017) found proximal concentrations of heavy metals in SWRO brine of 15±20%, 3.5±20%, 3±20%, 0.4±20%, and 0.13±20% µg/l for copper, chromium, nickel, iron, and molybdenum, respectively. The concentration values observed are significantly lower than the acute and chronically harmful levels for certain marine species, reported by S. Gheorghe et al. in their study on the toxic effects of metals in aquatic ecosystems (Gheorghe et al., 2017).

10.3.3 Cleaning Chemicals

RO filters must be cleaned every three to six months, but this depends on the quality of the water that goes into the system and how well the pretreatment works (Ihsanullah et al., 2021). How to clean the membrane relies on what kind of fouling it has. Acidic solutions (pH 2–3) break down metal oxides or scales, and alkaline solutions (pH 11–12) remove silt deposits and biofilms from membranes (Lee & Jepson, 2021). Cleaning products often have extra chemicals in them to help clean membranes better. These chemicals are called detergents (like dodecylsulfate and dodecylbenzene sulfonate) or oxidants (like sodium perborate and sodium hypochlorite), and they can harm aquatic life if they are dumped into the water without being treated first. Membranes are usually disinfected after they have been cleaned or before they are stored (Andersen & Bank, 2012). Biocides that do not oxidize, like formaldehyde, glutaraldehyde, or thiazole, can be used on the membrane instead of biocides that do oxidize, like chlorine and hydrogen peroxide. These biocides work very well and are dangerous to aquatic life if they enter the ocean. As part of thermal desalination, warm, acidic seawater is used to wash heat exchanger surfaces and remove alkaline scales (Lattemann & Höpner, 2008b). Chemical inhibitors like benzo triazole derivatives are often added to keep the plant from rusting. Since the inhibitors are not used in large amounts, they probably will not have any short-term toxic effects. However, both living and nonliving things break down this chemical reasonably slowly (Petersen, Frank, et al., 2018).

Kress et al. (2020) have provided comprehensive work on the specific implications of seawater desalination on the marine environment and aquatic life. N. Kress is a compilation of current observations on seawater desalination's impacts on the marine environment and aquatic life (Kress et al., 2018). These observations come from various data sources, including published literature, monitoring reports, environmental impact assessments (EIAs), and many other sources. Several factors that directly impact the marine environment, such as intake and outfall, are considered in this work. However, the scope of this analysis is limited to problems specific to the maritime environment; it does not consider other environmental impacts, such as gas emissions and others. To have a more in-depth conversation about the environmental impacts of desalination, it is essential to consider all of the components involved in the desalination process (Bar-Zeev et al., 2021).

The local population will feel the effects of desalination despite the method or technology used. Some excellent social, developmental, and economic effects include providing a steady water supply for residences, industries, and farms (Mezher et al., 2011). However, as we have already established, desalination has many undesirable environmental impacts (EIs), so finding ways to reduce or eliminate them requires extensive research. This is essential for the decision-making phase through the design, building, and operating phases (Al-Dousari et al., 2012).

The recovery ratio of SWRO desalination is typically 50–55%, while that of thermal desalination techniques is 30–40%. According to Lattemann and Höpner (2008a), this results in a saline discharge of approximately 50% to 70% of the input stream, which imposes various operational and EIs. The method of desalinating saltwater has undergone numerous process modifications. Nevertheless, the cost and higher energy demand prevents recovery rates from exceeding 55%. Because they have both reached their operating optimum levels, disposing of brine in the marine environment is their only option (Morillo et al., 2014). Giwa and Dindi (2017) provided an overview of the many approaches to brine management in their study, focusing on topics such as reduction and valuable applications for salt products. Muhammad Yaqub and Lee (2019) highlighted the growing energy requirement for a zero-liquid discharge (ZLD). It is commonly used in thermal procedures that require a lot of energy; therefore, it is essential to weigh the benefits of using it against the expenses (Giwa & Dindi, 2017).

A mix of distinct mesh screens and a moderate intake velocity is essential to minimize the effects of open intakes. By placing intakes away from productive areas, such as deeper waters, offshore, or underground (e.g., by using beach wells), it is possible to reduce the impingement and entrainment of larger organisms, such as fish or turtles, and minimize the entrainment of smaller plankton organisms, eggs, and larvae (Missimer et al., 2018). Given the generally superior quality of the intake water in these areas compared to nearshore and surface waters, it is often sufficient to do minimal or no chemical processing. However, the first soil disruption during subterranean intakes or extended pipe installation may be relatively more significant, mainly when it entails drilling or excavation operations. (Beach)Well intakes are exclusively suitable for plants of small or medium size. Consideration should be given to the co-location of desalination and power plants for more extensive facilities whenever feasible. Utilizing the cooling water from the power plant as feedwater for the desalination plant can decrease the overall intake water volume, therefore mitigating the adverse effects of entrainment and impingement, chemical usage, and construction and land use impacts (Kumar et al., 2022).

10.4 CONCLUSION

This chapter discusses the EIs of desalination plants, which have become increasingly important due to the global need for freshwater. While addressing water scarcity, desalination technology poses significant environmental and socio-economic challenges. The study indicates that most desalination occurs in areas such as the Mediterranean coast, the Gulf of Arabia, and the Red Sea, with numerous facilities globally. Greenhouse gas emissions from desalination plants and brine discharge, which impacts marine ecosystems and contains toxic chemicals, are two major environmental problems. The chapter additionally examines the socio-environmental consequences, including the inequitable distribution of freshwater and rising water costs. It underscores the necessity for improved governance and consideration of the socio-environmental drawbacks of desalination. Some studies conducted on the environmental effects of brine discharge are discussed in this chapter. These studies include the effects of brine discharge on marine life, salinity, water temperature, and dissolved oxygen levels. Additionally, it examines the possibility of heavy metal contamination and the utilization of chemicals in the early stages of the treatment process. The chapter concludes by stressing the importance of minimizing the environmental footprint of desalination through improved technologies and better management practices.

REFERENCES

Abdul-Wahab, S. A., & Abdo, J. (2007). Optimization of multistage flash desalination process by using a two-level factorial design. *Applied Thermal Engineering*, *27*(2–3), 413–421. https://doi.org/10.1016/j.applthermaleng.2006.07.010

Ahmed, M., & Anwar, R. (2012). An Assessment of the Environmental Impact of Brine Disposal in Marine Environment. *International Journal of Modern Engineering Research (IJMER)*, *2*(4), 2756–2761. www.ijmer.com

Al-Dousari, A., Al-Ghadban, A. N., & Sturchio, N. C. (2012). Marine environmental impacts of power-desalination plants in Kuwait. *Aquatic Ecosystem Health and Management, 15*, 50–55. https://doi.org/10.1080/14634988.2012.667336

Aleisa, E., & Al-Shayji, K. (2018). Ecological–economic modeling to optimize a desalination policy: case study of an arid rentier state. *Desalination*. https://doi.org/10.1016/j.desal.2017.12.049

Almasoudi, S., & Jamoussi, B. (2024). Desalination technologies and their environmental impacts: A review. *Sustainable Chemistry One World, 1* (November 2023), 100002. https://doi.org/10.1016/j.scowo.2024.100002

Al-Shayji, K., & Aleisa, E. (2018). Characterizing the fossil fuel impacts in water desalination plants in Kuwait: A Life Cycle Assessment approach. *Energy*. https://doi.org/10.1016/j.energy.2018.06.077

Ameen, F., Stagner, J. A., & Ting, D. S. K. (2018). The carbon footprint and environmental impact assessment of desalination. *International Journal of Environmental Studies*, *75*(1), 45–58. https://doi.org/10.1080/00207233.2017.1389567

Andersen, I., & Bank, W. (2012). Environmental Impacts of Desalination. *Renewable Energy Desalination: An Emerging Solution to Close the Water Gap in the Middle East and North Africa*, 111–126.

Aviram, R., Katz, D., & Shmueli, D. (2014). Desalination as a game-changer in transboundary hydro-politics. *Water Policy*. https://doi.org/10.2166/wp.2014.106

Bar-Zeev, E., Belkin, N., Speter, A., Reich, T., Geisler, E., & Rahav, E. (2021). Impacts of sewage outbursts on seawater reverse osmosis desalination. *Water Research*, *204*, 117631. https://doi.org/10.1016/j.watres.2021.117631

Bombar, G., Dölgen, D., & Alpaslan, M. N. (2016). Environmental impacts and impact mitigation plans for desalination facilities. *Desalination and Water Treatment*, *57*(25), 11528–11539. https://doi.org/10.1080/19443994.2015.1089198

California Coastal Commission. (2004). *California Coastal Commission 2004 - Google Scholar.* https://scholar.google.com/scholar?hl=en&as_sdt=0%2C5&q=California+Coastal+Commission+2004&btnG=

Chandrasekharam, D., Lashin, A., Al Arifi, N., & Al-Bassam, A. M. (2016). Red sea geothermal provinces. In *Red Sea Geothermal Provinces*. https://doi.org/10.1201/b20408

Chen, C., Sun, L., Liu, D., & Luo, X. L. (2019). Impact of Cumulative Fouling Characteristics on Full-cycle Operation Optimisation of Multi-effect Distillation Desalination System. *Chemical Engineering Transactions*, *76*(2017), 649–654. https://doi.org/10.3303/CET1976109

Danoun, R., & Jones, S. F. (2007). *Desalination Plants: Potential impacts of brine discharge on marine life*. https://core.ac.uk/download/pdf/41230762.pdf

Dawoud, M. A. (2012). Environmental Impacts of Seawater Desalination: Arabian Gulf Case Study. *International Journal of Environment and Sustainability*, *1*(3). https://doi.org/10.24102/ijes.v1i3.96

Elsaid, K., Kamil, M., Sayed, E. T., Abdelkareem, M. A., Wilberforce, T., & Olabi, A. (2020). Environmental impact of desalination technologies: A review. *Science of the Total Environment*, *748*. https://doi.org/10.1016/j.scitotenv.2020.141528

Elsaid, K., Sayed, E. T., Abdelkareem, M. A., Baroutaji, A., & Olabi, A. G. (2020a). Environmental impact of desalination processes: Mitigation and control strategies. *Science of the Total Environment*, *740*. https://doi.org/10.1016/j.scitotenv.2020.140125

Elsaid, K., Sayed, E. T., Abdelkareem, M. A., Baroutaji, A., & Olabi, A. G. (2020b). Environmental impact of desalination processes: Mitigation and control strategies. *Science of the Total Environment, 740*, 140125. https://doi.org/10.1016/j.scitotenv.2020.140125

Esrafilian, M., & Ahmadi, R. (2019). Energy, environmental and economic assessment of a polygeneration system of local desalination and CCHP. *Desalination*. https://www.sciencedirect.com/science/article/pii/S0011916418313377

Figler, A., B-Béres, V., Dobronoki, D., Márton, K., Alex Nagy, S., & Bácsi, I. (2019). Salt tolerance and desalination abilities of nine common green microalgae isolates. *Water*. https://doi.org/10.3390/w11122527

Frank, H., Fussmann, K., Rahav, E., Bar Zeev, B. (2019). Chronic effects of brine discharge from large-scale seawater reverse osmosis desalination facilities on benthic bacteria. *Water Research*. https://www.sciencedirect.com/science/article/pii/S0043135418310649

Fritzmann, C., Löwenberg, J., Wintgens, T., & Melin, T. (2007). State-of-the-art of reverse osmosis desalination. *Desalination, 216*(1–3), 1–76. https://doi.org/10.1016/j.desal.2006.12.009

Gao, L., Liu, G., Zamyadi, A., Wang, Q., & Li, M. (2021). Life-cycle cost analysis of a hybrid algae-based biological desalination–low pressure reverse osmosis system. *Water Research*. https://www.sciencedirect.com/science/article/pii/S004313542100155X

Ghaffour, N., Lattemann, S., Missimer, T., Choon Ng, K., & Amy, G. (2013). Seawater desalination using renewable energy: solar, geothermal, and wind 1 2. In *Proceeding of International Conference on Applied Energy (ICAE 2013)*.

Gheorghe, S., Stoica, C., Vasile, G. et al. (2017). Metals toxic effects in aquatic ecosystems: modulators of water quality. *Books.Google.Com*. https://books.google.com/books?hl=en&lr=&id=GMOQDwAAQBAJ&oi=fnd&pg=PA59&dq=Gheorghe+et+al.,+2017+&ots=d0ArRY3Si6&sig=ZuZ6Tgf24b6-AUf-E6hRHvvQuZ0

Giwa, A., & Dindi, A. (2017). An investigation of the feasibility of proposed solutions for water sustainability and security in water-stressed environment. *Journal of Cleaner Production, 165*, 721–733. https://doi.org/10.1016/j.jclepro.2017.07.120

Goh, P. S., Liang, Y. Y., & Ismail, A. F. (2021). Energy efficient seawater desalination: Strategies and opportunities. *Energy Technology, 9*(5), 1–26. https://doi.org/10.1002/ente.202100008

Heck, N., Lykkebo Petersen, K., Potts, D. C., Haddad, B., & Paytan, A. (2018). Predictors of coastal stakeholders' knowledge about seawater desalination impacts on marine ecosystems. *Science of the Total Environment, 639*, 785–792. https://doi.org/10.1016/j.scitotenv.2018.05.163

Hoepner, T. (1999). A procedure for environmental impact assessments (EIA) for seawater desalination plants. *Desalination, 124*(1–3), 1–12. https://doi.org/10.1016/S0011-9164(99)00083-1

Ihsanullah, I., Atieh, M. A., Sajid, M., & Nazal, M. K. (2021). Desalination and environment: A critical analysis of impacts, mitigation strategies, and greener desalination technologies. *Science of the Total Environment, 780*, 146585. https://doi.org/10.1016/j.scitotenv.2021.146585

Williams, J. (2022). Desalination in the 21st century: a critical review of trends and debates. *Water Alternatives*. https://www.water-alternatives.org/index.php/alldoc/articles/vol15/v15issue2/666-a15-2-8/file

Jones, E., Smakhtin, V., Qadir, M., Van Vliet, M. T. H., Kang, S.-M., & Qadir, M. (2019). The state of desalination and brine production: A global outlook. *Science of The Total Environment, 657*, 1343–1356. https://doi.org/10.1016/j.scitotenv.2018.12.076

Keirstad, R. E. (1962). Minerals for the chemical and allied industries (Johnstone, Sydney J.; Johnstone, Margery G.). *Journal of Chemical Education, 39*(12), A970. https://doi.org/10.1021/ed039pa970

Kennedy, A. D., & Jacoby, C. A. (1999). Biological indicators of marine environmental health: Meiofauna - A neglected benthic component? *Environmental Monitoring and Assessment, 54*(1), 47–68. https://doi.org/10.1023/A:1005854731889

Khawaji, A. D., Kutubkhanah, I. K., & Wie, J. M. (2007). A 13.3 MGD seawater RO desalination plant for Yanbu Industrial City. *Desalination, 203*(1–3), 176–188. https://doi.org/10.1016/j.desal.2006.02.018

Khawaji, A. D., Kutubkhanah, I. K., & Wie, J. M. (2008). Advances in seawater desalination technologies. *Desalination, 221*(1–3), 47–69. https://doi.org/10.1016/j.desal.2007.01.067

Kress, N. (2019). Marine impacts of seawater desalination: Science, management, and policy. In *Marine Impacts of Seawater Desalination: Science, Management, and Policy*. https://doi.org/10.1016/C2016-0-01997-6

Kress, N., Gertner, Y., & Shoham-Frider, E. (2020). Seawater quality at the brine discharge site from two mega size seawater reverse osmosis desalination plants in Israel (Eastern Mediterranean). *Water Research, 171*, 115402. https://doi.org/10.1016/j.watres.2019.115402

Kress, N., Oceanographic, I., & Galil, B. (2018). Impact of seawater desalination by reverse osmosis on the marine environment. In *Efficient Desalination by Reverse Osmosis Nurit Kress and Bella Galil Query Please provide accessed date for all web address. R. April.*

Kumar, V., Agrawal, S., Bhat, S. A., Américo-Pinheiro, J. H. P., Shahi, S. K., & Kumar, S. (2022). Environmental impact, health hazards, and plant-microbes synergism in remediation of emerging contaminants. *Cleaner Chemical Engineering, 2*, 100030. https://doi.org/10.1016/j.clce.2022.100030

Lattemann, S., & Höpner, T. (2008a). Environmental impact and impact assessment of seawater desalination. *Desalination, 220*(1–3), 1–15. https://doi.org/10.1016/j.desal.2007.03.009

Lattemann, S., & Höpner, T. (2008b). Environmental impact and impact assessment of seawater desalination. *Desalination, 220*(1–3), 1–15. https://doi.org/10.1016/j.desal.2007.03.009

Lattemann, S., & Höpner, T. (2008c). Impacts of seawater desalination plants on the marine environment of the Gulf. *Protecting the Gulf's Marine Ecosystems from Pollution*, 191–205. https://doi.org/10.1007/978-3-7643-7947-6_10

Lattemann, S., & Höpner, T. (2008d). Impacts of seawater desalination plants on the marine environment of the Gulf. *Protecting the Gulf's Marine Ecosystems from Pollution*, 191–205. https://doi.org/10.1007/978-3-7643-7947-6_10

Lee, K., & Jepson, W. (2021). Environmental impact of desalination: A systematic review of Life Cycle Assessment. *Desalination, 509*(October 2020), 115066. https://doi.org/10.1016/j.desal.2021.115066

Loganathan, K., Sulaiti, H., Bukhari, S. (2019). Distribution of salinity and trace elements in surface seawater of the Arabian Gulf surrounding the State of Qatar. *Desalination and Water Treatment*. https://www.sciencedirect.com/science/article/pii/S1944398624154768

Ma, Q., & Lu, H. (2011). Wind energy technologies integrated with desalination systems: Review and state-of-the-art. *Desalination, 277*(1–3), 274–280. https://doi.org/10.1016/j.desal.2011.04.041

Mauguin, G., & Corsin, P. (2005). Concentrate and other waste disposals from SWRO plants: characterization and reduction of their environmental impact. *Desalination*. https://www.sciencedirect.com/science/article/pii/S0011916405004510

Fragkou, M.C. (2018). Disclosing water inequalities at the household level under desalination water provision: The case of Antofagasta, Chile. In *Tapping the Oceans: Seawater Desalination and the Political Ecology of Water*. Edward Elgar Publishing. https://doi.org/10.4337/9781788113816.00006

McEvoy, J., & Wilder, M. (2012). Discourse and desalination: Potential impacts of proposed climate change adaptation interventions in the Arizona–Sonora border region. *Global Environmental Change, 22*(2), 353–363. https://doi.org/10.1016/J.GLOENVCHA.2011.11.001

Meerganz von Medeazza, G. L. (2005). "Direct" and socially-induced environmental impacts of desalination. *Desalination, 185*(1–3), 57–70. https://doi.org/10.1016/j.desal.2005.03.071

Mezher, T., Fath, H., Abbas, Z., & Khaled, A. (2011). Techno-economic assessment and environmental impacts of desalination technologies. *Desalination, 266*(1–3), 263–273. https://doi.org/10.1016/j.desal.2010.08.035

Missimer, T. M., Watson, I., Maliva, R. G., Ghaffour, N., Dehwah, A. H. A., Woolschlager, J., & Hegy, M. (2018). Impacts of natural pore-water and offshore aquifer chemistry on the operation and economics of some subsurface intakes types for SWRO plants. *Desalination and Water Treatment, 132*, 1–9. https://doi.org/10.5004/dwt.2018.23134

Morillo, J., Rosado, D., El, H., Lantania, B., Usero, J., El Bakouri, H., Riaza, A., & Bernaola, F.-J. (2014). Comparative study of brine management technologies for desalination plants. *Desalination*. https://doi.org/10.1016/j.desal.2013.12.038

Yaqub, M., & Lee, W. (2019). Zero-liquid discharge (ZLD) technology for resource recovery from wastewater: A review. *Science of the Total Environment, 681*, 551–563. https://doi.org/10.1016/j.scitotenv.2019.05.062

Neuparth, T., Costa, F., & Costa, M. (2002). Effects of Temperature and Salinity on Life History of the Marine Amphipod Gammarus locusta. Implications for Ecotoxicological Testing. *Ecotoxicology, 11*(1), 61–73. https://doi.org/10.1023/A:1013797130740

Omerspahic, M., Al-Jabri, H., Siddiqui, S. A., & Saadaoui, I. (2022). Characteristics of Desalination Brine and Its Impacts on Marine Chemistry and Health, With Emphasis on the Persian/Arabian Gulf: A Review. *Frontiers in Marine Science, 9*(April), 1–12. https://doi.org/10.3389/fmars.2022.845113

Panagopoulos, A. (2021). Water-energy nexus: desalination technologies and renewable energy sources. *Environmental Science and Pollution Research, 28*(17), 21009–21022. https://doi.org/10.1007/s11356-021-13332-8

Panagopoulos, A. (2022). Study and evaluation of the characteristics of saline wastewater (brine) produced by desalination and industrial plants. *Environmental Science and Pollution Research, 29*(16), 23736–23749. https://doi.org/10.1007/s11356-021-17694-x

Panagopoulos, A., & Haralambous, K. J. (2020). Environmental impacts of desalination and brine treatment - Challenges and mitigation measures. *Marine Pollution Bulletin, 161*. https://doi.org/10.1016/j.marpolbul.2020.111773

Petersen, K. L., Frank, H., Paytan, A., & Bar-Zeev, E. (2018). Impacts of Seawater Desalination on Coastal Environments. In *Sustainable Desalination Handbook: Plant Selection, Design and Implementation*, 437–463. https://doi.org/10.1016/B978-0-12-809240-8.00011-3

Petersen, K. L., Paytan, A., Rahav, E., Levy, O., Silverman, J., Barzel, O., Potts, D., & Bar-Zeev, E. (2018). Impact of brine and antiscalants on reef-building corals in the Gulf of Aqaba – Potential effects from desalination plants. *Water Research, 144*, 183–191. https://doi.org/10.1016/j.watres.2018.07.009

Haurwitz, R., Broad, T., Collins, J., Carman, N., Arroyo, J., Krishna, H. et al. (2008). *Desalination: Is It Worth Its Salt? A Primer on Brackish... - Google Scholar.* https://scholar.google.com/scholar?hl=en&as_sdt=0%2C5&q=DESALINATION%3A+IS+IT+WORTH+ITS+SALT%3F+A+Primer+on+Brackish+and+Seawater+Desalination&btnG=

Roberts, D. A., Johnston, E. L., & Knott, N. A. (2010). Impacts of desalination plant discharges on the marine environment: A critical review of published studies. *Water Research, 44*(18), 5117–5128. https://doi.org/10.1016/j.watres.2010.04.036

Rybar, S., Vodnar, M., Vartolomei, L., León Méndez, R., Blas, J., & Ruano, L. (2005). Experience with renewable energy source and SWRO desalination in Gran Canaria. *Fgwater.Com*, 5–100. https://www.fgwater.com/static/upload/file/20230615/1686811256486630.pdf

Sadhwani, J. J., Veza, J. M., & Santana, C. (2005). Case studies on environmental impact of seawater desalination. *Desalination, 185*(1–3), 1–8. https://doi.org/10.1016/j.desal.2005.02.072

Safrai, I., & Zask, A. (2008). Reverse osmosis desalination plants - marine environmentalist regulator point of view. *Desalination*, *220*(1–3), 72–84. https://doi.org/10.1016/j.desal.2007.01.023

Sharaf, M. A., Nafey, A. S., & García-Rodríguez, L. (2011). Exergy and thermo-economic analyses of a combined solar organic cycle with multi effect distillation (MED) desalination process. *Desalination*, *272*(1–3), 135–147. https://doi.org/10.1016/j.desal.2011.01.006

Shenvi, S. S., Isloor, A. M., & Ismail, A. F. (2015). A review on RO membrane technology: Developments and challenges. *Desalination*, *368*, 10–26. https://doi.org/10.1016/j.desal.2014.12.042

Sirota, R., Winters, G., Levy, O., Marques, J., Paytan, A., Silverman, J., Sisma-Ventura, G., Rahav, E., Antler, G., & Bar-Zeev, E. (2024). Impacts of Desalination Brine Discharge on Benthic Ecosystems. *Environmental Science and Technology*, *58*(13), 5631–5645. https://doi.org/10.1021/acs.est.3c07748

Sola, I., Fernández-Torquemada, Y., Forcada, A., Valle, C., del Pilar-Ruso, Y., González-Correa, J. M., & Sánchez-Lizaso, J. L. (2020a). Sustainable desalination: Long-term monitoring of brine discharge in the marine environment. *Marine Pollution Bulletin*, *161*, 1–30. https://doi.org/10.1016/j.marpolbul.2020.111813

Sola, I., Fernández-Torquemada, Y., Forcada, A., Valle, C., del Pilar-Ruso, Y., González-Correa, J. M., & Sánchez-Lizaso, J. L. (2020b). Sustainable desalination: Long-term monitoring of brine discharge in the marine environment. *Marine Pollution Bulletin*, *161*. https://doi.org/10.1016/j.marpolbul.2020.111813

Sun, S., Zhou, X., Liu, H., Jiang, Y., Zhou, H., Zhang, C., et al. (2021). Unraveling the effect of inter-basin water transfer on reducing water scarcity and its inequality in China. *Water Research*. https://www.sciencedirect.com/science/article/pii/S0043135421001299

Swyngedouw, E., & Williams, J. (2016). From Spain's hydro-deadlock to the desalination fix. *Water International*. https://doi.org/10.1080/02508060.2016.1107705

Tularam, G., & Ilahee, M. (2007). Initial loss estimates for tropical catchments of Australia. *Environmental Impact Assessment Review*. https://doi.org/10.1016/j.eiar.2006.12.006

Voutchkov, N. (2020). Introduction to Desalination Project Design and Delivery. *Sun Cam*.

Voutchkov, N., Kaiser, G., et al. (2019). Sustainable Management of Desalination Plant Concentrate-Desalination Industry Position Paper–Energy and Environment Committee of the International Desalination Association .... *Researchgate. Net*. https://www.researchgate.net/profile/Nikolay-Voutchkov/publication/336879311_SUSTAINABLE_MANAGEMENT_OF_DESALINATION_PLANT_CONCENTRATE_DESALINATION_INDUSTRY_POSITION_PAPER_ENERGY_AND_ENVIRONMENT_COMMITTEE_OF_THE_INTERNATIONAL_DESALINATION_ASSOCIATION_IDA_Presenter/links/6052f66ca6fdccbfeae956ab/SUSTAINABLE-MANAGEMENT-OF-DESALINATION-PLANT-CONCENTRATE-DESALINATION-INDUSTRY-POSITION-PAPER-ENERGY-AND-ENVIRONMENT-COMMITTEE-OF-THE-INTERNATIONAL-DESALINATION-ASSOCIATION-IDA-Presenter.pdf

Wilder, M. O., Aguilar-Barajas, I., Pineda-Pablos, N., Varady, R. G., Megdal, S. B., McEvoy, J., Merideth, R., Zúñiga-Terán, A. A., & Scott, C. A. (2016). Desalination and water security in the US–Mexico border region: assessing the social, environmental and political impacts. *Water International*, *41*(5), 756–775. https://doi.org/10.1080/02508060.2016.1166416

Zhou, D., Zhu, L., Fu, Y., Zhu, M., Xue, L. (2017). Development of lower cost seawater desalination processes using nanofiltration technologies—A review. *Desalination*. https://www.sciencedirect.com/science/article/pii/S0011916415300564

Zhou, J., Chang, V. W. C., & Fane, A. G. (2011). Environmental life cycle assessment of reverse osmosis desalination: The influence of different life cycle impact assessment methods on the characterization results. *Desalination*, *283*, 227–236. https://doi.org/10.1016/j.desal.2011.04.066

Index

Pages in **bold** refer to tables.

A

B

C

D

E

G

H